MATERIALS RESEARCH SOCIETY
SYMPOSIUM PROCEEDINGS VOLUME 1394

# Oxide Semiconductors-Defects, Growth and Device Fabrication

November 28 – December 3, 2011
Boston, Massachusetts, USA

Printed from e-media with permission by:

Curran Associates, Inc.
57 Morehouse Lane
Red Hook, NY 12571
www.proceedings.com

ISBN: 978-1-62748-214-1

Some format issues inherent in the e-media version may also appear in this print version.

©Materials Research Society 2011

This reprint is produced with the permission of the Materials
Research Society and Cambridge University Press.

This publication is in copyright, subject to statutory exception and to the
provisions of relevant collective licensing agreements. No reproduction
of any part may take place without the written permission of Cambridge
University Press.

Cambridge University Press
Cambridge, New York, Melbourne, Madrid, Cape Town,
Singapore, São Paulo, Delhi, Tokyo, Mexico City

Cambridge University Press
32 Avenue of the Americas, New York, NY 10013-2473, USA
www.cambridge.org

Materials Research Society
506 Keystone Drive, Warrendale, PA 15086
www.mrs.org

CODEN: MRSPDH

ISBN: 978-1-62748-214-1

Cambridge University Press has no responsibility for the persistence or
accuracy of URLs for external or third-part Internet Web sites referred to
in this publication and does not guarantee that any content on such Web sites
is, or will remain, accurate or appropriate.

**Additional copies of this publication are available from:**

Curran Associates, Inc.
57 Morehouse Lane
Red Hook, NY 12571 USA
Phone:  845-758-0400
Fax:     845-758-2634
Email:  curran@proceedings.com
Web:    www.proceedings.com

# Oxide Semiconductors-Defects, Growth and Device Fabrication

Materials Research Society Symposium Proceedings
Volume 1394

Boston, Massachusetts, USA
28 November - 3 December 2011

# TABLE OF CONTENTS

**Photoluminescence Due to Group IV Impurities in ZnO** ........................................ 1
*J. Cullen, K. Johnston, M. Henry, E. McGlynn*

**Emergence Of Blue Emission With Decreasing Film Thickness And Grain Size for ZnO Grown Via Thermal Oxidation Of Zn-Metal Films** ........................ 7
*L. Covington, R. Stansell, J. Moore*

**Room Temperature Ferromagnetism Of Fe-doped ZnO and MgO Thin Films Prepared By Ink-Jet Printing** ............................................................... 13
*M. Fang, W. Voit, A. Kyndiah, Y. Wu, L. Belova, K. Rao*

**Experimental Evidence for Nitrogen as a Deep Acceptor in ZnO** ...................... 21
*M. Tarun, M. Iqbal, M. McCluskey, J. Huso, L. Bergman*

**Nitrogen Pair □ Hydrogen Complexes In ZnO and P-Type Doping** ................... 27
*A. Boonchun, W. Lambrecht, J. T-Thienprasert, S. Limpijumnong*

**Growth Techniques for Bulk ZnO and Related Compounds** ............................ 32
*D. Klimm, D. Schulz, S. Ganschow, Z. Galazka, R. Uecker*

**Microstructural, Optical and Electrical Properties Of Post-Annealed ZnO:Al Thin Films** ............................................................................... 42
*C. Charpentier, P. Prod'Homme, L. Francke, P. Cabarrocas*

**$ZnS_xO_{1-x}$ Films Grown on Flexible Substrates** ........................................ 48
*J. Huso, H. Che, J. Morrison, D. Thapa, M. Huso, S. Rhodes, B. Blanchard, W. Yeh, M. McCluskey, L. Bergman*

**Intrinsic Paramagnetic Defects in Zirconium and Hafnium Oxide Films** ......... 53
*R. Schwartz, H. Muller, P. Adams, J. Barrie, R. Lacoe*

**Effects of Hydrogen Ion Implantation on Structural Properties of Silver Implantation in ZnO Crystals** ..................................................................... 61
*F. Yaqoob, M. Huang*

**Influence Of Post-Deposition Annealing On Structural, Optical And Electrical Characteristics Of NiO/ZnO Thin Film Hetsero-Junction** ............... 68
*M. Tyagi, M. Tomar, V. Gupta*

**Anomalous Diffusion of Intrinsic Defects in $K^+$ Implanted ZnO using Li as Tracer** ................................................................................................. 75
*L. Vines, P. Veuvonen, A. Kuznetsov, J. Wong-Leung, C. Jagadish, B. Svensson*

**Improved Resistive Switching Properties in $HfO_2$-based ReRAMs by Hf/Au Doping** ................................................................................................... 81
*X. He, N. Tokranova, W. Wang, R. Geer*

**Heteroepitaxial Growth of ZnO Films on $Gd_3Ga_5O_{12}$ Garnet Substrates** ...... 87
*Y. Ono, H. Matsui, H. Tabata*

**ZnO Coated Nanoparticle Phosphors** ........................................................... 93
*M. Kobayashi*

**Diffusion Of Ion Implanted Indium And Silver In ZnO Crystals**.................................. 101
*F. Yaqoob, M. Huang*

**Effect of Low Power Deposition and Low Oxidation Temperature on the Interfacial and Structural Properties of sputtered HfO$_2$ Gate Dielectrics**.................. 108
*A. Minko, G. Belo, S. Rudenja, D. Buchanan*

**Dependence of Annealing Temperature on the Conductivity Changes of ZnO and MgZnO Nanoparticle Thin Films from Annealing in a Hydrogen Atmosphere at Mild Temperatures**.................................. 114
*C. Berven, L. Sanchez, S. Chava, H. Young, J. Dick, J. Morrison, J. Huso, L. Bergman*

**Fabrication Of Titanium Oxide Film With High Crystallinity By The New Electrochemical Techniques** ....................................... 120
*H. Ishizaki, S. Ito*

**Effects of Substrate Pre-deposition Annealing and Deposition Parameters on the Properties of RF Sputter-deposited ZnO Films** .................................. 127
*T. Oder, M. McMaster, A. Smith, N. Velpukonda, D. Sternagle*

**Author Index**

Mater. Res. Soc. Symp. Proc. Vol. 1394 © 2012 Materials Research Society
DOI: 10.1557/opl.2012.214

## Photoluminescence due to Group IV impurities in ZnO

J. Cullen[1], K. Johnston[2,3], M. O. Henry[1,2] and E. McGlynn[1]
[1]School of Physical Sciences, Dublin City University, Collins Avenue, Dublin 9, Ireland
[2]ISOLDE Collaboration, CERN, CH-1211 Geneva 23, Switzerland
[3]Technische Physik, Universitat des Saarlandes, D66041 Saarbrucken, Germany

### ABSTRACT

We report the results of photoluminescence measurements on ZnO bulk crystals implanted with both stable and radioactive species involving the group IV impurities Ge, Si and Sn. We previously confirmed the identity of a line emerging at 3.3225 eV as being related to Ge and present here uniaxial stress data which show that the defect responsible has trigonal symmetry. Experiments with Si provide circumstantial evidence of a connection with the well-known line at 3.333 eV. Our measurements indicate that for the case of Sn on the Zn site luminescence is not observed. We also confirm that the $I_9$ and $I_2$ lines are due to substitutional In impurities.

### INTRODUCTION

Steady progress has been made in recent years in unambiguously identifying the principal band edge bound exciton ($I_n$) lines in the photoluminescence (PL) spectra of ZnO, and several of these lines have now been assigned to bound exciton recombination at either neutral or ionised group III donor impurities on the Zn site. Nevertheless, several of the dominant $I_n$ lines remain to be positively identified, in addition to some new lines recently reported [1, 2] in the PL spectra below the usual bound exciton region. These new lines are sharp, with full width at half maximum values similar to those of the $I_n$ lines. In contrast to the $I_n$ lines, however, their thermal binding energy is much lower than their spectral binding energy [2].

Ion implantation allows for controlled doping of impurities and for unparalleled control over the depth of these impurities after doping. However, the implantation process can induce defects not native to the semiconductor which may survive annealing procedures, making spectral features that appear after ion implantation difficult to assign unambiguously to the implanted impurity. Radioisotopes have been used in conjunction with traditional spectroscopic techniques for decades [3] and their use removes some of the ambiguity by correlating the decaying (increasing) intensity of a feature to the mother (daughter) isotope in the sample. Radioisotopes have already been combined with PL to accurately study the behaviour of specific dopants in a variety of semiconductors, including ZnO [3,4].

In this paper we expand on earlier work [2] that used radioisotopes to identify a sharp PL line at 3.3225 eV with Ge impurities. We provide more detail on the properties of this Ge-related line, and we also examine samples implanted with Si and with radioactive Ag that decays through In to stable Sn. We also note results obtained by other workers for the case of the group IV impurity Pb [5].

## EXPERIMENT

For radiotracer ion implantation high quality single crystal ZnO obtained from Tokyo Denpa Ltd. (Tokyo, Japan) and Rubicon Technology, Inc. (Illinois, USA) was used. Radioactive ion implantation was performed at the ISOLDE facility at CERN using an energy of 60 keV and typical doses of $5 \times 10^{12}$ atoms $cm^{-2}$. Details of the radiotracer implantation procedure are given elsewhere [2]. For stable isotope implantation an energy of 100 keV and typical doses of $1 \times 10^{13}$ atoms $cm^{-2}$ were used. The use of different implantation conditions is due to instrument constraints. However, the range/straggle for the radioactive and stable implantations are 32/14 nm and 49/20 nm, respectively [6], resulting in approximately the same peak concentrations sufficiently close to the sample surfaces. Following implantation, samples were annealed in $O_2$ gas or in air at 750°C for 30 minutes. We have found over many ZnO PL studies that the use of air or $O_2$ as annealing ambient has no significant effect on the band edge PL.

For uniaxial stress measurements, hydrothermally grown single crystal samples (supplied by Tokyo Denpa Ltd.) with planes cut perpendicular to the (11-20), (10-10) and (0001) axes were used in order to investigate the lines under stresses parallel and perpendicular to the c-axis (0001). Details of the stress apparatus are given elsewhere [7].

Luminescence was generated by the 325-nm line of a HeCd laser operating in the range 80 - 200 mW. The spectra were recorded at temperatures in the range 2.7 - 10 K using Janis closed cycle helium cryostats. The luminescence was analysed by a SPEX 0.75 m grating spectrometer equipped with a $LN_2$-cooled Jobin-Yvon CCD detector for the radiotracer implanted samples, and a Jobin-Yvon iHR320 grating spectrometer fitted with an Andor Newton EM-CCD detector cooled to -25°C for the uniaxial stress and normal PL measurements.

## RESULTS AND DISCUSSION

A typical PL spectrum of as-received Tokyo Denpa ZnO is shown in figure 1 (a) below.

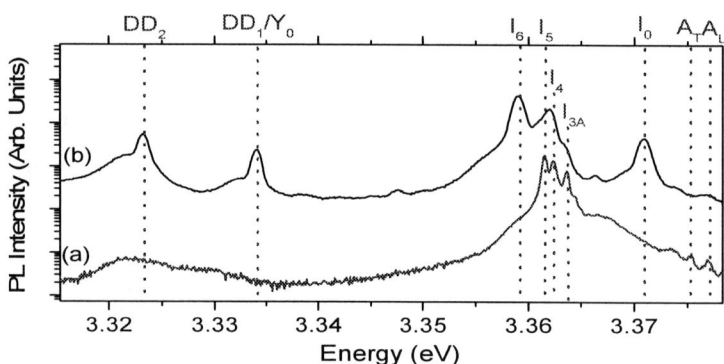

**Figure 1.** (a) A medium resolution PL spectrum for c-plane as-received ZnO. Shown in (b) is the same sample after ion implantation with Ge impurities and subsequent annealing. Spectra are vertically shifted for clarity and were recorded at 10 K.

The spectrum is composed of the free A-exciton (transverse and longitudinal) just below 3.38 eV, the ionised and neutral donor bound exciton lines around 3.36 – 3.37 eV and the lower energy two-electron satellites around 3.31 to 3.33 eV. Not shown here are the longitudinal optical phonon replicas which occur at ~73 meV intervals extending down to lower energies. The recently investigated 'Y-Lines' [1] that occur in the vicinity of 3.333 eV are not observed in any of our as-received hydrothermally grown material supplied by Tokyo Denpa.

## ZnO:Ge

The PL spectrum of a sample implanted with stable Ge and subsequently annealed at 750°C in $O_2$ for 30 minutes is shown in figure 1 (b) above. Two sharp lines, at 3.3225 eV and 3.333 eV, appear in the spectrum well below the normal $I_n$ line region; the relative intensities of the I lines are also changed from those in the as-received material. The 3.333 eV line (variously labelled as $Y_0$ or $DD_1$) which also occurs in un-implanted but annealed samples, has been reported previously [1,2,8] and in an earlier work we reported on the main properties of the 3.3225 eV line, including a positive identification of the line with Ge impurities on the Zn site [2]. We now present the results of uniaxial stress and polarisation measurements on this line, which we label $DD_2$ hereafter. Regarding the polarisation properties, our results show that the line is not significantly polarised, with a difference of ~19% between the intensities for polarisation parallel and perpendicular to the c-axis. This compares to a difference of 55% for the I lines.

Representative PL spectra under uniaxial stress are presented in figure 2. For stress parallel to the c-axis all lines shift to higher energy, with similar (but unequal) shift rates, while they all shift to lower energies for stress perpendicular to the c-axis ( data not shown), also with similar but unequal shift rates.

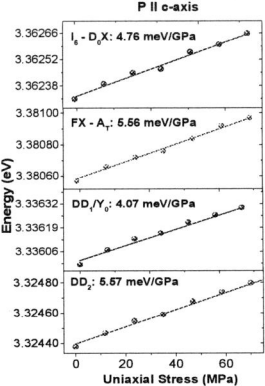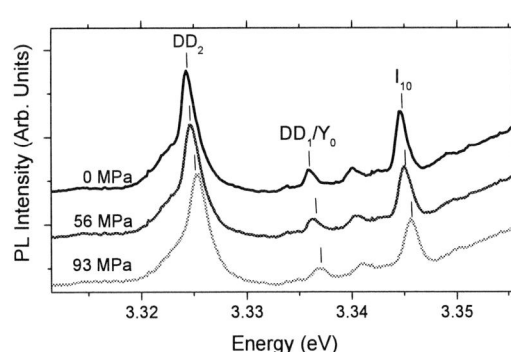

**Figure 2.** Spectra and shift rates for the $DD_2$ and nearby lines under uniaxial stress parallel to the c-axis. All lines shift to higher energy for stress parallel to the c-axis, with similar shift rate.

It is noteworthy that the $DD_2$ line does not split for any stress, and up to the largest stress value used, the line shape remains essentially the same. The absence of any splitting of this line for the two stresses means that the line originates at a centre of trigonal symmetry [9] further supporting the assignment of the line to isolated Ge atoms, and therefore corroborating a previous identification of the line with Ge on Zn sites [2]. The observed shift rate for the free exciton under stress parallel to the c-axis of + 5.56 meV/GPa is in reasonable agreement with the value of + 4.71 meV/GPa reported by Wagner *et al* [1] in a recent study of the $Y_0$ line.

The contrast between the large spectral binding energy relative to the free exciton position (~57 meV) and the observed weak thermal binding energy (~15 meV) was noted in previous work [2], and the tentative identification of the line with exciton recombination at Ge double donors in that work remains to be proven.

### ZnO:Sn

We explored the case of ZnO:Sn by means of a sample implanted with a radioactive pre-cursor that transforms to stable Sn as its final decay product. Radioactive $^{117}$Ag is available at ISOLDE at high purity thanks to a highly selective laser ionisation process. The decay sequence of this isotope is Ag(73 s )/Cd(2.5 hrs)/In(43 mins)/Sn(stable). The short half life of $^{117}$Ag rules out the possibility of detecting Ag-related signals in PL, but the populations of Cd, In and Sn over a 24-hour period allow for the examination of In-related $D_0X$ and $D_+X$ lines in addition to the possibility of observing Cd- or Sn-related PL.

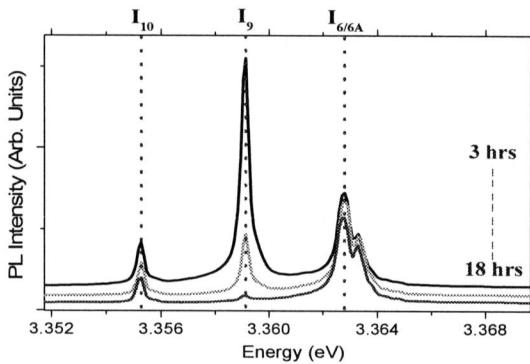

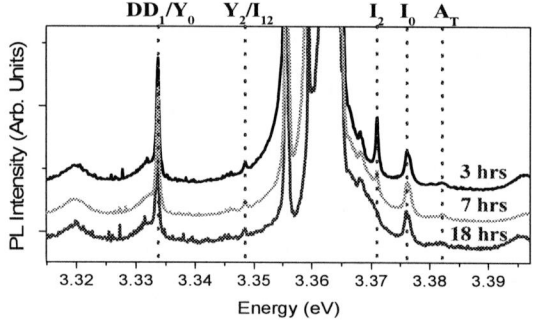

**Figure 3.** Representative spectra showing the disappearance of the $I_9$ and $I_2$ features due to substitutional In on a Zn site ($In_{Zn}$) as the In decays to Sn (with a half life 43.2 mins). Spectra are vertically shifted for clarity and shown on a linear y-axis.

The PL data over five half-lives of [117]In are shown in figure 3 above, and fits to the time dependence of the line intensities are shown in figure 4 below. It is immediately clear that both the $I_9$ and $I_2$ lines are due to In: the calculated half life of $43\pm2$ minutes is in very good agreement with the tabulated half life of 43.2 minutes. The association of $I_9$ with In had been confirmed earlier using [111]In [10], but this is the first proof of $I_2$ being the $D_+X$ line for In donors. There is no evidence of Sn-related (or Cd-related) PL in the data we recorded during this experiment.

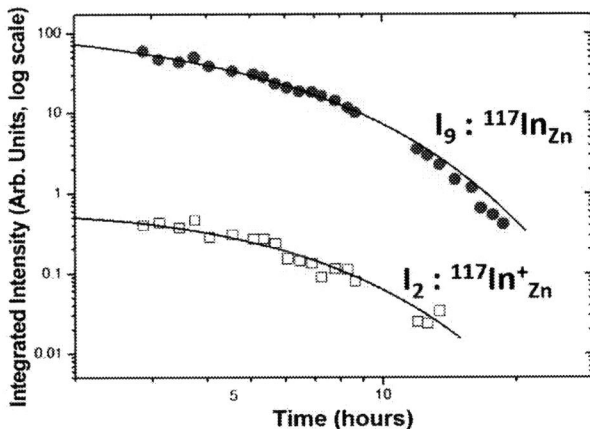

**Figure 4.** The intensity values over time represented by the red circles and blue squares for $I_9$ and $I_2$, respectively, and the accompanying fits to the decay of the lines.

## ZnO:Si

We examined samples implanted with stable Si, but did not observe any PL lines unique to the case of ZnO:Si samples. However, we found some indications that the line at 3.333 eV ($Y_0$ or $DD_1$) may be related to Si, as its intensity was slightly enhanced in ZnO:Si compared to as-received samples. We note that Si is one of the principal impurities in the Tokyo Denpa material we used. Nevertheless, at this stage we cannot make a positive association of the 3.333 eV line with Si.

## SUMMARY AND CONCLUSIONS

We have undertaken a PL study of several of the group IV impurities in ZnO, and of these, we find clear evidence with considerable spectral detail only for the case of ZnO:Ge. The uniaxial stress data in particular show that the $DD_2$ line at 3.3225 eV originates at a centre with trigonal symmetry, adding to the evidence that this line is due to a simple single-atom impurity in the ZnO crystal. Overall, the data we present here corroborate the evidence from earlier work linking the line to $Ge_{Zn}$ impurities. It remains to be proven, however, that the source of the luminescence is bound exciton recombination at Ge double donor centres.

For the case of ZnO:Si we have circumstantial evidence only of a link between Si and the 3.333 eV ($Y_0$) line. So, although Si in ZnO has been predicted to act in a similar fashion to Ge [11], the data we have obtained to date do not provide proof of this. It should be noted, however, that Si is a major impurity in the ZnO crystals we have used, and future work will be carried out on starting material with lower Si contamination.

We now consider the cases of ZnO:Sn and ZnO:Pb. We find that samples with Sn impurities on Zn sites, formed by the decay of radioactive In, do not produce any Sn-related PL. Other workers, in a study of nanostructured ZnO likely to include significant Pb contamination, suspected a correlation of $I_7$ with Pb, but examination of Pb-implanted bulk material in the same study did not corroborate the result [5]. However, subsequent work by the same group indicates that there may be a Pb-related PL line at 3.3637 eV [12].

It is clear from the foregoing discussion that there remain several questions regarding the properties of group IV impurities in ZnO. To date, the only cases to present evidence of band edge recombination are Ge and Pb. Whether any of the other group IV impurities also act as binding centres for exciton recombination remains to be established.

## ACKNOWLEDGEMENTS

This work was supported by the German BMBF under Contract No. 05KK7TS2 and by Science Foundation Ireland under Contract No. 08/RFP/PHY1558. This work has also been supported by the European Community as an Integrating Activity 'Support of Public and Industrial Research using Ion-beam Technology (SPIRIT)' under EC contract no. 227012.

## REFERENCES

1. M. R. Wagner, G. Callsen, J. S. Reparaz, J.-H. Schulze, R. Kirste, M. Cobet, I. A. Ostapenko, S. Rodt, C. Nenstiel, M. Kaiser, A. Hoffmann, A. V. Rodina, M. R. Phillips, S. Lautenschlager, S.Eisermann and B. K. Meyer, *Phys. Rev. B* **84**, 035313 (2011).
2. K. Johnston, J. Cullen, M. O. Henry, E. McGlynn and M. Stachura, *Phys. Rev. B* **83**, 125205 (2011).
3. M. Deicher and the ISOLDE Collaboration, *Physica B* **389**, 51–57 (2007).
4. M. O. Henry, M. Deicher, R. Magerle, E. McGlynn and A. Stotzler, *Hyperfine Interactions* **129**, 443 (2000).
5. R. J. Mendelsberg, J. V. Kennedy, S. M. Durbin and R. J. Reeves, *J. Vac. Sci. Technol. B* **27**, 3 (2009).
6. http:www.surrey.ac.uk/ati/bc/research/modelling_simulation/suspre.htm
7. C. O'Morain, K. G. McGuigan, M. O. Henry and J. D. Campion, *Meas. Sci. Tech.* **3**, 337 (1992).
8. A. Schildknecht, R. Sauer and K. Thonke, *Physica B* **205**, 340 (2003).
9. E. McGlynn and M. O. Henry, *Phys. Rev. B* **76**, 184109 (2007).
10. S. Muller, D. Stichtenoth, M. Uhrmacher, H. Hofsass, C. Ronning and J. Roder, *Appl. Phys. Lett.* **90**, 012107 (2007).
11. J. L. Lyons, A. Janotti and C. G. Van de Walle, *Phys. Rev. B* **80**, 205113 (2009).
12. R. J. Mendelsberg, M. W. Allen,, S. M. Durbin and R. J. Reeves, *Phys. Rev. B* **83**, 205202 (2011).

Mater. Res. Soc. Symp. Proc. Vol. 1394 © 2012 Materials Research Society
DOI: 10.1557/opl.2012.245

# Emergence of blue emission with decreasing film thickness and grain size for ZnO grown via thermal oxidation of Zn-metal films

L.R. Covington, R. Stansell, and J.C. Moore[*]
Department of Chemistry and Physics, Coastal Carolina University, Conway, SC 29528, U.S.A

## ABSTRACT

We have investigated the photoluminescence properties of ZnO grown on sapphire substrates via the thermal oxidation of Zn-metal films at various temperatures and thicknesses. X-ray diffraction (XRD) spectra indicate that the resulting films possess a polycrystalline hexagonal wurtzite structure without preferred orientation. For films having a thickness of 200 nm, crystal grain size was observed to decrease with increasing annealing temperature up to 600°C, and then increase at higher temperatures. Thicker films demonstrated a modest increase in grain size as temperature increased from 300°C to 1200°C. The influence of film thickness on the optical properties was investigated using room temperature photoluminescence (PL). Specifically, PL spectra indicate four emission bands: excitonic ultraviolet, blue, and deep-level green and yellow emission. The strongest UV emission and narrowest full width at half maximum (0.09 eV) was observed for films having a thickness of 200 nm and annealed at low temperature (300°C). As film thickness decreased, we observed the emergence of blue emission. The emergence of blue emission when depletion width grows relative to the bulk suggests that the origin of the blue emission is related to the negatively charged Zinc interstitials found within the deletion region near the interface.

## INTRODUCTION

Zinc oxide (ZnO) is a wide bandgap semiconductor that has attracted a great deal of attention with demonstrated applications in ultraviolet (UV) light detection, air-quality monitoring, missile warning systems, gas detection, and utilization as light-emitting diodes. Thin films of ZnO have been grown via a variety of methods, including the sol-gel method, molecular beam epitaxy, chemical vapor deposition, and the thermal oxidation of Zn-metal films.[1,2,3] Specifically with regards to thermal oxidation, Cho et al. demonstrate growth of undoped ZnO films having a polycrystalline wurtzite structure, where grain size is seen to increase with annealing temperature.[4] The photoluminescence (PL) spectrum for ZnO typically includes a broad ultraviolet (UV) band emission attributed to exitonic binding energies and green and yellow band emission, due to various defect states.[5,6,7] In several instances, blue-band emission has also been reported.[8,9,10,11]

Wu et al. argue that deep-level green and yellow emission corresponds to a recombination of a delocalized electron close to the conduction band with a deeply trapped hole in the single ionized oxygen vacancy and the single negatively charged interstitial oxygen ion centers, respectively.[5] There has been much debate on the cause of defects leading to specific PL emission, and several possibilities have been suggested such as growth environment, annealing temperature, and film thickness. Wang et al. report weak deep level PL emission bands

---

[*] Corresponding author, email: moorejc@coastal.edu, phone: 843-349-2985

compared to UV emission, and green band intensity that increases at higher temperatures when Zn-metal films are oxidized in air.[12] However, Chen et al. show decreasing deep level band intensities with increasing temperature when films are oxidized in oxygen ambient via a two-step process.[13] These seemingly conflicting reports suggest that the oxidizing agent and final annealing temperature is critical in determining PL behavior.

The origin of blue emission is controversial. Lee et al. describe strong blue emission possibly caused by stress resulting from the volume expansion of the ZnO transformed from Zn during high treatment temperature.[14] Whereas Zhang et al. have shown that the intensity of the blue PL peak is strongly dependent on the oxygen pressure.[8] They conclude that one source of blue level emission is from the electron transition from the shallow donor of oxygen vacancies to the valence band, while another is electron transmission from the shallow donor level of zinc interstitials to the valence band. However, Wu et al. argue that blue emission is related solely to Zn interstitials found within the depletion region. They argue that a low ratio of bulk to depletion region results in blue emission, since bulk-related defects associated with deep-level emission would dominate except when bulk volume is comparable to that of the depletion region.[5]

We investigate the effects of annealing temperature, film thickness, and grain size on PL emission for ZnO grown via thermal oxidation of Zn-metal films. Specifically, if green and blue emission result from defect-related energy levels in the bulk and depletion region, respectively, as suggested by Wu et al., and we tailor the bulk-to-depletion-region ratio via decreases in thickness and grain size, then we should observe a corresponding decrease in the green-to-blue emission ratio.

## EXPERIMENT

In this study, Zinc films were grown on $c$-plane sapphire substrates using direct current sputter deposition without reactive gas.[15] The sputter cathode used was a 1" diameter 99.99% purity zinc target mounted on a water-cooled stage. A turbomolecular pump maintained a background pressure of $10^{-6}$ mbars before deposition. During deposition, an argon pressure of approximately $10^{-2}$ mbars was maintained via a metal leak valve and pump throttling. Deposition times ranged from 10-60 minutes at sputter power between 15-30 W. All resulting Zn films were initially oxidized by thermal annealing in an open air muffle furnace at a temperature of 300°C for over 24 hours. Some films where then re-annealed at temperatures of 600°C, 900°C, and 1200°C for two hours. After thermal oxidation, all films where removed from the furnace and allowed to cool in air ambient.[16,17,18]

Structural properties of the resulting films where investigated using x-ray diffraction (XRD). XRD spectra indicate that after annealing, the resulting ZnO films possess a polycrystalline hexagonal wurtzite structure without preferred orientation. Grain size was calculated from the Scherrer formula and XRD spectra, and confirmed via atomic force microscopy (AFM, Anfatec Level). X-ray analysis is discussed in more detail in ref. 15. Surface morphology and film thickness was measured using AFM. Film thickness was confirmed via reflectometry using a broad-spectrum fluorescent source and a UV-Vis spectrophotometer. Post-annealed ZnO films demonstrated thicknesses ranging from ~200 nm to ~600 nm. Bandgap energies were determined by absorption band edge using the same spectrophotometer with an integrated tungsten-deuterium source. Photoluminescence spectra where acquired at room temperature using a HeCd laser (325 nm) as an excitation source at a power $P = 0.3$ W/cm$^2$.

**DISCUSSION**

Grain size was characterized by both AFM and XRD. For films having thicknesses of 400 nm and 600 nm, grain size was observed to increase with increasing annealing temperature from 300°C to 1200°C. Interestingly, for 200 nm thick films, grain size decreased with increasing temperature up to a certain point. AFM measurements show a decrease in particle diameter from approximately 150 nm to approximately 100 nm at annealing temperatures of 300°C and 600°C, respectively, with no further significant change in size as temperature was further increased. Grain size, as determined by the full width at half maximum in the XRD spectra (not shown), was also found to decrease with increasing temperature up to 600°C, with relatively small increases at higher temperatures, which is consistent with AFM measurements. This observation appears inconsistent with some reports in the literature for thermally oxidized ZnO films.[4,12] This discrepancy may be the result of differences in studied temperature regimes, the variations in film thickness, and/or our two-step thermal annealing process, where metallic zinc films are all initially oxidized at low temperature. Furthermore, our metallic zinc films display a significantly different texture and larger initial particle size in comparison to films grown via other methods, which has been shown to affect resulting film morphology and structure.[19]

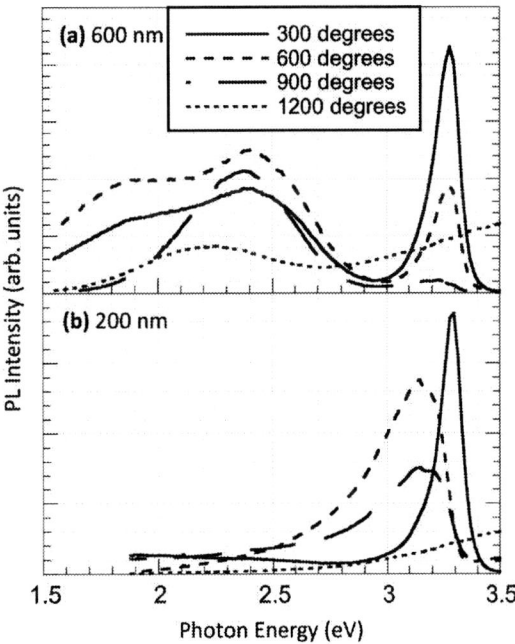

**Figure 1:** PL spectra of ZnO films grown at 300°C, 600°C, 900°C and 1200°C at thicknesses of (a) 600 nm and (b) 200 nm. For thicker films, increasing green band emission relative to UV emission is seen with increasing temperature. Thinner films demonstrate little green band emission; however, a significant redshift in UV emission and asymmetrical band broadening is observed at higher temperatures.

Figure 1 shows the PL spectra of ZnO films with thickness of (a) 600 nm and (b) 200 nm thermally annealed at various temperatures. The spectra for the 600 nm thick films all show asymmetric broad bands in the yellow-green region, with no significant yellow-green emission observed for the 200 nm thick films. For all thicknesses, a more narrow band in the UV is observed, which is associated with excitonic emission; however, the 200 nm film demonstrates an asymmetric and broad band in the blue-UV region at higher temperatures. For all thicknesses, the strongest UV excitonic emission is observed at low temperature, with decreasing UV emission observed with increasing annealing temperature. Interestingly, for thinner films, increasing temperature results in a significant redshift (0.15 eV) in the UV excitonic peak and an asymmetrical peak broadening.

To determine whether changes in the bandgap were responsible for the observed red shift in UV emission, we measured bandgap energies for varying annealing temperature. Figure 2 shows the Tauc plot obtained from the absorption spectra for thin films annealed at 300°C and 600°C.[20] Absorption measurements indicate an indirect bandgap that redshifts from 3.25 eV to 3.10 eV with increasing annealing temperature. No further redshift in bandgap was observed with increasing temperature past 600°C (not shown). This shift in bandgap energy is consistent with the observed shift in UV emission peak in the PL spectra shown in fig. 1(b), and consistent with bandgap shifts with decreasing film thickness reported in the literature.[21] Jain et al. speculate that this red shift is the result of an increase in interstitial zinc atoms. Wu et al. and Dijken et al. both demonstrate that an increase in particle size should result in a redshift in energies, which appears inconsistent with our results, since in this temperature and thickness regime, we see a decrease in grain size.[5,22] However, these studies discuss systems where quantum size effects become relevant, and the particle sizes in this study are sufficiently large such that the shift in bandgap can not be explained via a similar mechanism.

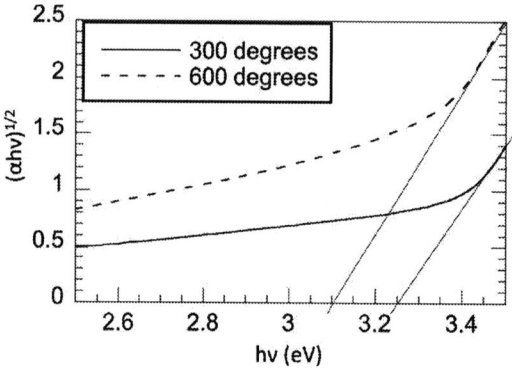

**Figure 2:** Tauc plot obtained from absorption measurements for 200 nm thick ZnO films annealed at 300°C and 600°C. Indirect bandgap energies are obtained from the y-intercept of the extrapolation for the linearly increasing region. Bandgap energies are observed to redshift approximately 0.15 eV with increasing temperature, which is consistent with observed shifts in UV excitonic emission.

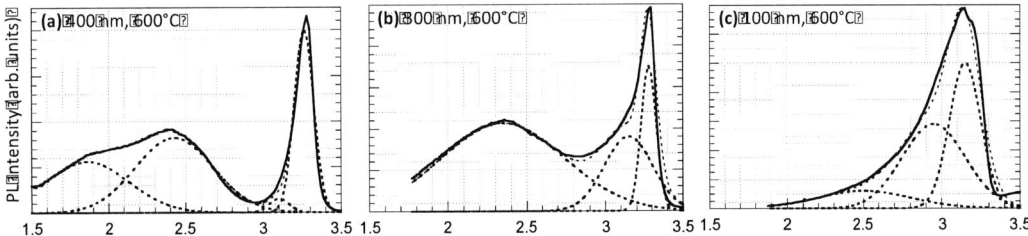

**Figure 3:** PL spectra of ZnO films annealed at 600°C and grown to thicknesses of approximately (a) 600 nm, (b) 400 nm, and (c) 200 nm. Solid lines indicate measured PL emission. Dashed lines correspond to fits derived from linear combination of Gaussian peaks centered

Figure 3 shows PL spectra (solid lines) for ZnO films annealed at 600°C having three thicknesses: approximately (a) 600 nm, (b) 400 nm, and (c) 200 nm. To determine the affect of film thickness on the blue emission, we modeled overall PL emission as the linear combination of four Gaussian peaks centered on yellow, green, blue, and UV energies (dashed lines). The UV excitonic emission is observed as expected considering the bandgap energies of the films. Specifically, UV and blue emissions for the 200 nm film are redshifted in a manner consistent with the observed shift in bandgap

## RESULTS

The presence of blue emission explains the observed asymmetric shape of the dominant blue-UV peak for all thicknesses. The ratio of green-to-blue emission decreases with decreasing film thickness. The PL spectra for the 200 nm thick film exhibited the most dramatic blue band emission and very low green band emission. It has been suggested that blue emission results from zinc interstitials found in the depletion region, so an approximately 400 nm blue emission should only be observed in a sample with a wide depletion region relative to the bulk. Otherwise, deep-level green emission will dominate.[9,10] If blue emission is related to zinc interstitials, then we would observe the emergence of blue emission with decreasing film thickness, since limiting thickness lowers the ratio of bulk to depletion region.

Temperature and grain size may also contribute, since no blue emission is observed for 200 nm films annealed at 300°C [see fig. 1(b)]. As discussed, we observe larger grain size via XRD and AFM for 200 nm films annealing at 300°C, resulting in a larger bulk to depletion region ratio, which could contribute to weaker blue emission. As with the green emission, reaction kinetics associated with temperature may also contribute, were thinner films would have shorter diffusion paths for reactive oxygen species during oxidation, and lower temperatures would slow the $Zn/O_2$ reaction. The result could be fewer zinc interstitials, which would manifest as weaker blue emission and a red shift in both blue and UV emission, as is observed.

## ACKNOWLEDGMENTS

The authors would like to thank Everett Carpenter and Mikhail Reshchikov from Virginia Commonwealth University for providing access to their XRD and PL systems, respectively. This work was funded by NSF DMR \#1104600.

## REFERENCES

[1] U. Özgür, Y.I. Alivov, C. Liu, A. Teke, M.A. Reshchikov, S. Dogan, V. Avrutin, S.J. Cho, and H.Morkoç, J. Appl. Phys. 98, 041301 (2005).

[2] J.C. Moore, S.M. Kenny, C.S. Baird, H. Morkoç, A.A. Baski, J. Appl. Phys. 105, 116102 (2009).

[3] S. Chevtchenko, J.C. Moore, Ü. Özgür, X. Gu, B. Nemeth, J.E. Nause, A.A. Baski, H. Morkoç, Appl. Phys. Lett., 89, 182111 (2006).

[4] S. Cho, J. Ma, Y. Kim, Y. Sun, G. Wong, and J. Ketterson, Appl. Phys. Lett. 75(18), 2761 (1999).

[5] X. Wu, G. Siu, C. Fu, and H. Ong, Appl. Phys. Lett. 78, 2285 (2001).

[6] Z. Wanga, X. Zu, S. Zhu, and L. Wang, Physica E 35(1), □199–202 (2006).

[7] Z. Fu, C. Guo, B. Lin, and G. Liao, Chin. Phys. Lett. 15(6), □457–459 (1998).

[8] D. Zhang, Z. Xue, and Q. Wang, J. Phys. D 35, 2837 □(2002).

[9] L. Zhao, J. S. Lian, Y. H. Liu, and Q. Jiang, Transactions □of Nonferrous Metals Society of China 18(1), 145–149 □(2008).

[10] J. Zhao, L. Hu, Z. Wang, Y. Zhao, X. Liang, and M. Wang, □Appl. Surf. Sci. 229(1-4), 311–315 (2004).

[11] S. Fujihara, Y. Ogawa, and A. Kasai, Chem. Mater. 16(2965) (2004).

[12] Y. Wang, S. Lau, H. Lee, S. Yu, B. Tay, X. Zhang, and □H. Hng, J. Appl. Phys. 94, 354 (2003).

[13] S. Chen, Y. Liu, J. M, D. Zhao, Z. Zhi, Y. Lu, J. Zhang, □D. Shen, and X. Fan, J. Cryst. Growth 240(3-4), 467–472 □(2002).

[14] M. Lee and H. Tu, Jap. J. Appl. Phys. 47(2), 980–982 □(2008).

[15] J.C. Moore, L.R. Covington, R.L. Foster, E.J. Gee, M.R. Jones, S.A. Morris, Proc. of SPIE, 7940, 79401L (2011).

[16] V. Avrutin, Ü. Özgür, N. Izyumskaya, S. Chevtchenko, J. Leach, J.C. Moore, A.A. Baski, C. Litton, H.O. Everitt, K.T. Tsen, M. Abouzaid, P. Ruterana, and H. Morkoç, Proc. of SPIE 6474, 64741M (2007).

[17] V. Avrutin, Ü. Özgür, N. Izyumskaya, S. Chevtchenko, J. Leach, J.C. Moore, A.A. Baski, H.O. Everitt, K.T. Tsen, P. Ruterana, H. Morkoç, Mater. Res. Soc. Symp. Proc., 963E (2006).

[18] J.C. Moore, J.L Skrobiszewski, A.A. Baski. J. Vac. Sci. Technol. 25, 4, 812-815 (2007).

[19] R. Gupta, N. Shridhar, and M. Katiyar, Mater. Sci. Semi. □Proc. 5, 11 (2002).

[20] J. Tauc, Mater. Res. Bulletin 3(1), 37–46 (1968).

[21] A. Jain, P. Sagar, and R. M. Mehra, Mater. Sci. Pol. 25(1), □233–242 (2007).

[22] A. van Dijken, E. Meulenkamp, D. Vanmaekelbergh, and □A. Meijerink, J. Lumin. 90, 123–128 (2000).

**Mater. Res. Soc. Symp. Proc. Vol. 1394 © 2012 Materials Research Society**
DOI: 10.1557/opl.2012.824

# Room temperature ferromagnetism of Fe-doped ZnO and MgO thin films prepared by ink-jet printing

Mei Fang[1], Wolfgang Voit[1], Adrica Kyndiah[1], Yan Wu[2], Lyubov Belova[1] and K. V. Rao[1]

[1] Department of Materials Science and Engineering, KTH-Royal Institute of Technology, Stockholm, SE10044, Sweden.
[2] Faculty of Materials Science and Chemical Engineering, China University of Geosciences, Wuhan, 430074, P.R.China.

## ABSTRACT

Room temperature magnetic properties of un-doped, as well as 10 at.% Fe-doped ZnO and MgO single-pass layer of ink-jet printed thin films have been investigated to obtain insight into the role of the band gaps and mechanisms for the origin of ferromagnetic order in these materials. It is found that on doping with Fe, the saturation magnetization is enhanced by several-fold in both systems when compared with the respective un-doped thin films. For a ~28 nm thick film of Fe-doped ZnO (Diluted Magnetic Semiconductor, DMS) we observe an enhanced moment of $0.465\mu_B$ /Fe atom while it is around $0.111\mu_B$/Fe atom for the doped MgO (Diluted Magnetic Insulator, DMI) film of comparable thickness. Also, the pure ZnO is far more ferromagnetic than pure MgO at comparable low film thicknesses which can be attributed to defect induced magnetism originating from cat-ion vacancies. However, the film thickness dependence of the magnetization and the defect concentrations are found to be significantly different in the two systems so that a comparison of the magnetism becomes more complex for thicker films.

## INTRODUCTION

Diluted magnetic semiconducting oxides (DMS) and diluted magnetic insulating oxides (DMI) have attracted many experimental and theoretical interests in the recent few years [1-5]. By doping with transition metals (TM), room temperature ferromagnetism (RTFM) can be obtained in semiconductors and insulators thus invoking the possibilities to engineer new band gap tailored materials with functional electrical properties. It is well accepted that ferromagnetism originates from the unpaired electrons localized in partly filled $d$- sub-shells in transition elements. However, in recent years this explanation is challenged by the discovery of the so-called $d^0$ ferromagnetism, in which RTFM can be obtained even without partially filled $d$- or $f$- sub-shells in atoms. Materials like irradiate graphite [6], doped nonmagnetic oxides [7], and even un-doped oxides [8], e.g. ZnO [9,10] and MgO [11,12] thin films, have been found to exhibit robust ferromagnetism at and well above room temperature. The magnetism in these materials is not due to any partially filled $d$-orbital but arises from moments induced at the defect sites in the lattice which changes the band structure (e.g., spin splitting). In addition, free carriers in oxides can also contribute to RTFM [13,14]. Therefore, the effects of the donors, the charge transfer and the structure defects on the observed ferromagnetism should be considered in DMS and DMI materials. A direct comparison of magnetization between TM-doped and un-doped oxides can be a way to investigate the effect of doping and reveal the contribution of each mechanism in DMS and DMI materials.

It is well known that for TM-doped ZnO thin films the observed RTFM depends on the fabrication methods and conditions [15]. These observations can be subject to possible contaminations and process dependent factors. Ink-jet printing is a cost-effective, non-contact, contamination free, digital and maskless patterning approach to deposit films, and is an ideal process for developing electrical devices with the potential for large area manufacturing [16,17]. We present here a study of room temperature ferromagnetism in un-doped ZnO, MgO, and 10 at.% Fe-doped ZnO, MgO thin films prepared by depositing the precursor solutions with piezoelectric actuated ink-jet printheads. The origins of the observed RTFM in these thin films are discussed.

## EXPERIMENT

Precursor ink preparation: i) For un-doped ZnO thin films, zinc acetate dehydrate ($Zn(OOCCH_3)_2 \cdot 2H_2O$ from Alfa Aesar) was dissolved into 2-isopropoxyethanol (IPE, $(CH_3)_2CHO(CH_2)_2OH$ from SIGMA-ALDRICH) to form a cat-ion molar concentration of 0.25M. ii) For 10 at.% Fe doped-ZnO thin films, iron (II) acetate anhydrous ($Fe(OOCCH_3)_2$ from Alfa Aesar) was added to form the same cat-ion molar concentration with [Fe]:[Zn] = 1:9. Drops of ethanolamine ($NH_2(CH_2)_2OH$ from Alfa Aesar) were added to stabilize the precursor inks for printing ZnO and Fe-doped ZnO thin films. iii) For printing un-doped MgO thin films, magnesium acetate tetrahydrate ($Mg(OOCCH_3)_2 \cdot 4H_2O$ from Alfa Aesar) was used to form 0.25 M [Mg] in methoxyethanol (MOE, $CH_3O(CH_2)_2OH$, ACS 99.3% from Alfa Aesar). iv) For printing 10 at.% Fe-doped MgO thin films, the ink was prepared from $Mg(OOCCH_3)_2 \cdot 4H_2O$ and $Fe(OOCCH_3)_2$ with a molar ratio of [Fe]:[Mg] = 1:9 and a cat-ion molar concentration of 0.25 M in MOE. For magnesium based inks, drops of acetate acid (glacial 99+% from Alfa Aesar) were added to increase the solubility as well as the stabilization. The prepared inks were characterized by thermogravimetric analysis (TG, Perkin-Elmer TGS-2) for the best processing conditions after printing. The heating rate for TG analysis was 20 °C/min.

Ink-jet printing: Piezoelectric Xaar 126 drop-on-demand (DOD) ink-jet printheads were utilized in a custom-built printing station with heatable xy-stage to deposit thin films on glass and silicon substrates. The substrates were ultrasonic cleaned in acetone for 10 minutes followed by another 10 minutes in ethanol, and then soaked in isopropanol. After cleaning, the substrates were dried with a nitrogen gun and then pre-heated to 80 °C prior to printing. The printed films were dried on a hotplate at 150°C for 10 minutes to evaporate the solvents. The first complete layer we designate as the single-pass layer. Subsequently a sequence of layers were deposited to obtain thicker films. For the preparation of multi-pass printed films, each pass was printed after drying appropriately the previous layer. For comparing the properties of the two oxides we present the studies on single-pass deposited films.

Heat treatments: The dried films were annealed in air to obtain oxides. ZnO and Fe-doped ZnO thin films were prepared by annealing at 450 °C for 1 hour, while MgO and Fe-doped MgO thin films were obtained by annealing at 450°C for 2 hours and then post-annealing at 600 °C for 2 more hours, to achieve the maximum magnetization according to our previous studies [17,18].

Characterizations of the prepared films: Copper anode X-ray diffractometer (XRD, Siemens D5000) was used to study the phase and the structure of the films. Grazing incidence diffraction at 2° were collected in range of 2θ from 20° to 80° with step of 0.01° and speed of 8 seconds/step. A dual beam focused ion beam scanning electron microscope (FIB/SEM, FEI Nova 600 Nanolab) was used to obtain the top morphologies and the cross section images of the films. Room temperature magnetic hysteresis loops were determined by a superconducting quantum interference device (SQUID, Quantum Design MPMS2).

## RESULTS AND DISCUSSION

Figure 1 shows the thermogravimetric analysis of the precursors used for preparing Fe-doped thin films. The temperature depended weight changes of the as-prepared inks were first determined, and for both inks the major weight loss occurred at ~150 °C which corresponded to the evaporation of the solvents. (Boiling point for IPE is 142-144 °C and for MOE is 124-125 °C.) To monitor the weight changes of the acetates in the precursors, the inks were dried on a hotplate at 150°C for 10 minutes to evaporate the solvents before the TG analysis. For the precursors of Fe-doped ZnO, the major weight loss occurred between 220 and 440°C, which can be attributed to the decomposition of the acetates. According the molar mass (M), the weight percent for Fe-doped ZnO is calculated to be:

$$(\frac{M_{ZnO}}{M_{Zincacetate}} \times 0.9 + \frac{M_{FeO}}{M_{Iron(II)acetate}} \times 0.1) \times 100\% = 44\% \tag{1}$$

This is close to the obtained value from the TG analysis.

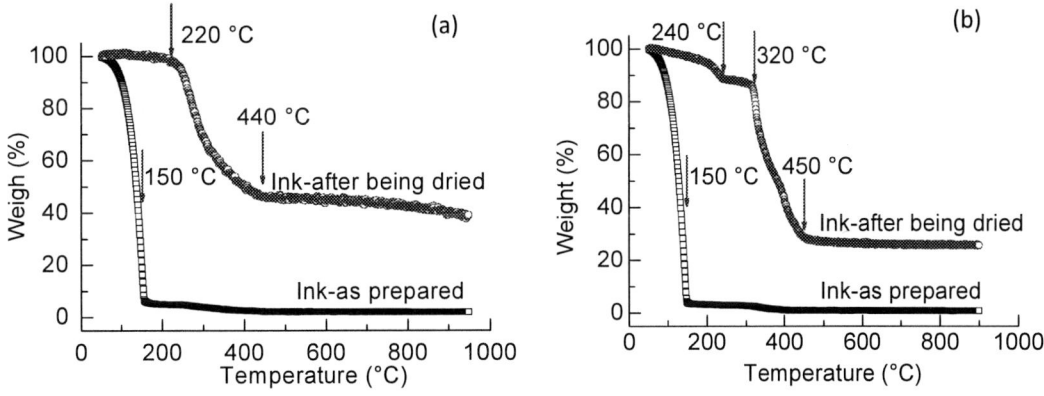

**Figure 1.** *Thermogravimetric analysis of precursor inks for preparation of (a) Fe-doped ZnO and (b) Fe-doped MgO thin films.*

For Fe-doped MgO (Fig.1b), the continuous weight loss until 240 °C corresponds to the loss of water molecules in $Mg(OOCCH_3)_2 \cdot 4H_2O$. The anhydrous magnesium acetate was stable until 320 °C, followed by the thermal decomposition in the temperature range between 320 and 450°C, during which the weight percent changed from 88.3% to 27.6%. According to the thermal decomposition of $Mg(OOCCH_3)_2$ and $Fe(OOCCH_3)_2$, the weight percent for Fe-doped MgO from anhydrous acetates is:

$$(\frac{M_{MgO}}{M_{Magnesiumacetate}} \times 0.9 + \frac{M_{FeO}}{M_{Iron(II)acetate}} \times 0.1) \times 100\% = 29.6\% \qquad (2)$$

Figure 2 shows the XRD intensity patterns for the thin films prepared on glass substrates. Standard ZnO and MgO diffraction peaks from Joint Committee on Power Diffraction Standards (JCPDS) No. 00-036-1451 and JCPDS No.00-045-0946 are also shown in the figures. The doped Fe atoms have different effects on the crystal structure of ZnO and MgO: For Fe-doped ZnO thin film the weaker intensity of diffraction peaks indicate poorer crystalline structure than for un-doped ZnO thin film (Fig. 2a), while for Fe-doped MgO thin films the crystal structure is strongly enhanced as compared to the almost amorphous un-doped MgO thin film (Fig. 2b). The lattice constants have been refined by using the Celref3 program according to the peak positions and are listed in Table I.

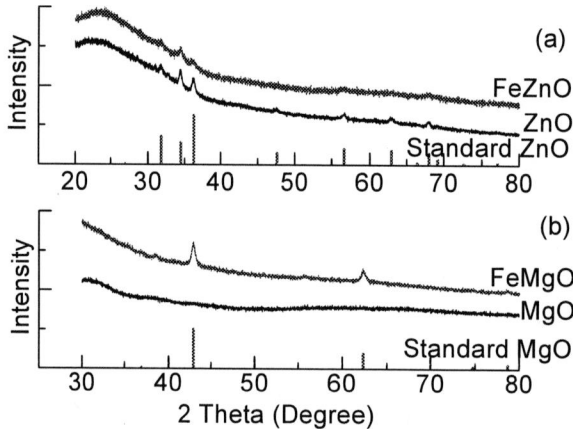

**Figure 2.** *XRD intensity patterns for ZnO, MgO and their Fe-doped thin films. The intensity data has been shifted along the Y-axis for clarity.*

**Table I.** Lattice constants of crystal thin films comparing with standard values from JCPDS.

| Samples | Space group | Lattice constants (nm) | Volume (nm$^3$) |
|---|---|---|---|
| Standard ZnO | P63mc | a=0.32498, c=0.52066 | 0.04762 |
| ZnO | P63mc | a=0.32527, c=0.52195 | 0.04782 |
| Fe-doped ZnO | P63mc | a=0.32566, c=0.52182 | 0.04793 |
| Standard MgO | Fm3m | a=0.42112 | 0.07468 |
| Fe-doped MgO | Fm3m | a=0.42128 | 0.07477 |

The prepared ZnO thin film has larger lattice constants than the standard values from bulk material, indicating the lattice strain in the thin film materials. The doped Fe atoms substitute Zn/Mg in ZnO/MgO structure, concomitantly the crystal lattice expands corresponding to the ionic radii [19] as well as the induced defects in the lattice by doping.

Figure 3 shows the FIB-SEM cross section images of single-layer printed thin films and their surface morphologies. The effect of Fe-doping on the crystal properties can be clearly seen from the images of the two systems. In the ZnO system, the porous structure of un-doped ZnO thin film becomes condensed and the grain size is reduced when doped with Fe, resulting in a thinner Fe-doped ZnO thin film. In the MgO system one can see the improvements of the crystalline structure from almost amorphous in un-doped MgO thin film into nano-crystalline grains in Fe-doped MgO thin film, which is consistent with the XRD analysis. Moreover, the uniformity of the thin film (see the thickness variation in the cross section image) improved considerably by doping Fe in the MgO thin film. Because of the dependence on the density (or the porous structure) and the crystal structure, pure MgO thin film is thicker than Fe-doped MgO thin film which has similar thickness as Fe-doped ZnO thin film.

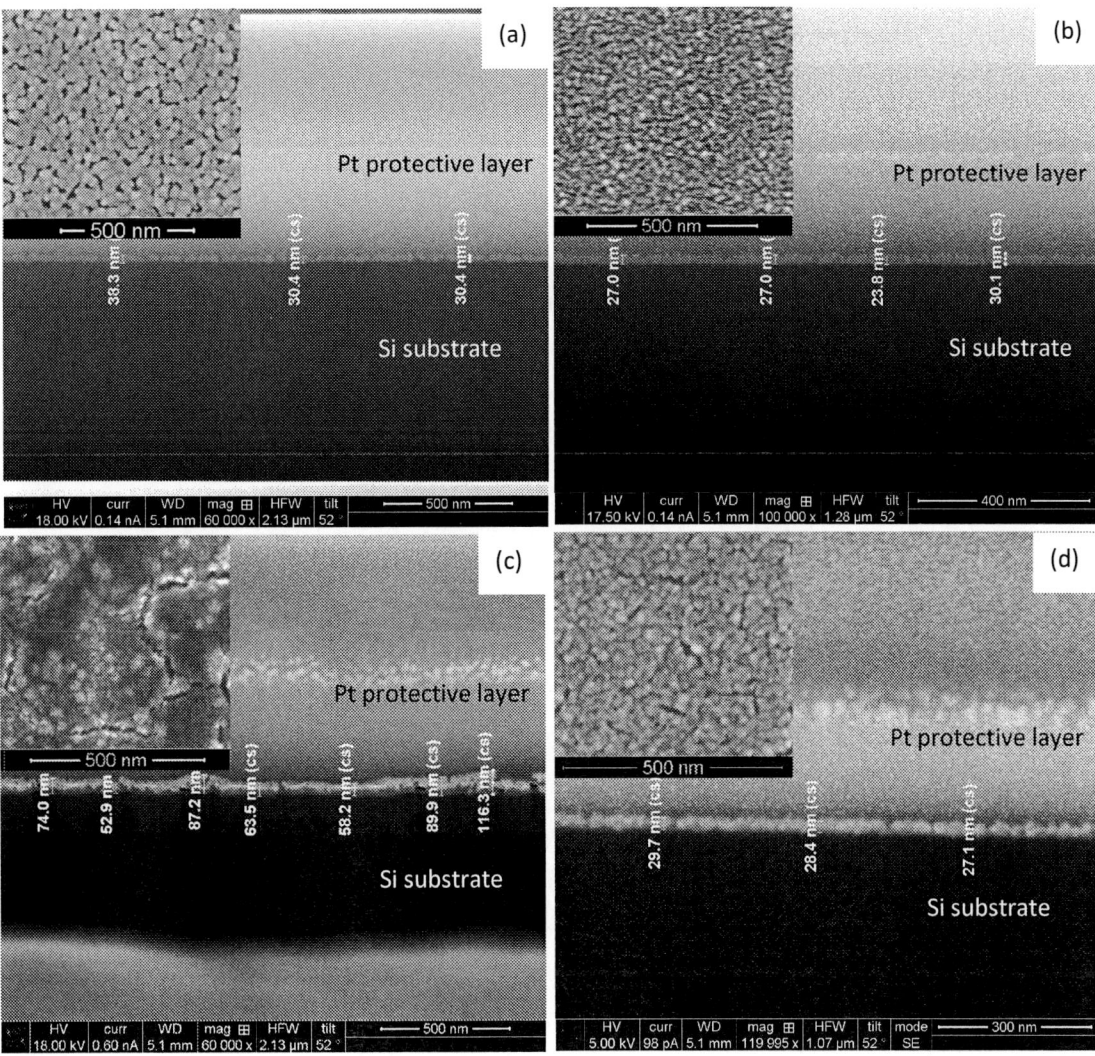

***Figure 3.*** *FIB-SEM cross section images of (a) ZnO, (b) Fe-doped ZnO, (c) MgO and (d) Fe-doped MgO thin films and their surface morphologies inserted in each corner.*

We suppose the doped Fe can increase the nucleation in both the systems, therefore crystalline ZnO decreases grain size in Fe-doped ZnO thin films while amorphous MgO becomes crystalline in Fe-doped MgO thin films. From the energy-dispersive X-ray spectroscopy (EDX) for thick films (printed with multi-pass), the ratio of [Fe] to [Zn]+[Fe] in Fe-doped ZnO thin films and [Fe] to [Fe]+[Mg] in Fe-doped MgO thin films ranges between 7 and 20 at.%, depending on the number of printed passes (film thickness) and the signal of elements from the substrate. Luckily, the doped MgO and ZnO have comparable film thickness!

Room temperature magnetic hysteresis loops of the prepared thin films are shown in Fig. 4. From the zoomed data of the loops around origin (the insets in figure 4(a,b)), the remanence and the coercivity ($H_C$, in the range of 20~80 Oe) can be observed, which is typical of the data in literature for robust but very weak ferromagnetism reported for oxide thin films [12,18,20,21]. We note that for both the systems, the Fe-doping enhanced the saturation magnetization (M) several-fold, i.e ~4 times in the ZnO system and ~8 times in the MgO system. The effective enhancement in magnetization (by 16.1 emu•cm$^3$ for a ~28 nm doped ZnO DMS single-printed layer, and by 5.5 emu•cm$^3$ for a ~28 nm doped MgO DMI single-printed layer) arises from the $Fe^{2+}$ state hybridized with cat-ion vacancy implies $p$-$d$ exchange coupling. We calculated the enhanced magnetization per Fe-atom (using a nominal 10 at.% Fe concentration) in 1 cm$^3$ Fe-doped thin films and the results are listed in Table II. It is found that in Fe-doped ZnO thin film the enhanced M is about $0.465\mu_B$ per Fe atom, while it is $0.11\mu_B$ per Fe atom in Fe-doped MgO thin film. This indicates that Fe-doping in ZnO system is more effective than in MgO system, which is consistent with the reduced exchange due to the wider band gap in MgO.

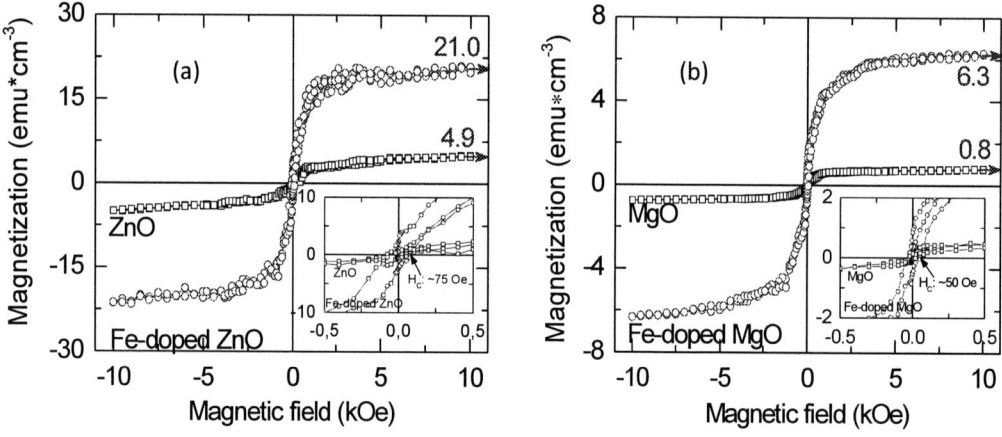

***Figure 4.*** *Room temperature magnetic hysteresis loops of prepared thin films. The insert figures show the zoom in of the loops around the origin, in which the remanence and finite coercivities can be observed.*

**Table II.** The calculated information of Fe atoms in 1 cm³ Fe-doped thin films.

| In 1 cm³ thin films | Unit cell volume (nm³) | Number of Fe atoms | Enhanced M (emu•cm⁻³) | Magnetization (μ_B per Fe atom) |
|---|---|---|---|---|
| Fe in Fe-doped ZnO | 0.04793 | $3.736 \times 10^{21}$ | 16.1 | 0.465 |
| Fe in Fe-doped MgO | 0.07477 | $5.350 \times 10^{21}$ | 5.5 | 0.111 |

Note: 1 $\mu_B$ = 9.274×10⁻²¹ emu (Bohr magneton).

## SUMMARY

In summary, by comparing the magnetic properties of un-doped ZnO and MgO , and their Fe-doped ink-jet printed thin films, the effects of crystal structure, band gap and Fe doping have been analyzed. The several-fold enhanced magnetization on Fe-doping in both the systems suggests that the magnetic properties of the films of the thickness considered here are related to the $3d$ electrons from Fe, as well as the defect induced consequences in both the un-doped and doped oxides. However, it is known that the magnetic properties in both the un-doped oxides are film thickness dependent and different from each other in details [12,17,20]. Thus while our comparison of the magnetism in 10 at% Fe doped ZnO and MgO are valid for the ~28 nm films, we point out that the relative comparison of the properties for thicker films is complex and is dependent on the film thickness, the defect structure and the quality of the films.

## ACKNOWLEDGMENTS

It is a pleasure to thank Prof. Gillian Gehring for many discussions on our results and studies. This project has been supported by a grant from the Swedish Agency VINNOVA. Mei Fang acknowledges the Chinese Scholarship Council for her PhD study. AK would like to thank the department of materials physics at KTH for a visiting opportunity and a grant to work on the project.

## REFERENCES

1. H. Ohno, Science **281,** 951-956 (1998).
2. A. Bonanni and T. Dietl, Chemical Society Reviews **39,** 528-539 (2010).
3. P. Sharma, A. Gupta, K. V. Rao, F. J. Owens, R. Sharma, R. Ahuja, J. M. O. Guillen, B. Johansson, and G. A. Gehring, Nature Materials **2,** 673-677 (2003).
4. S. Ramachandran, J. Narayan, and J. T. Prater, Applied Physics Letters **90,** 115321 (2007).
5. J. M. D. Coey, M. Venkatesan, and C. B. Fitzgerald, Nat Mater **4,** 173-179 (2005).
6. J. M. D. Coey, Solid State Sciences **7,** 660-667 (2005).
7. K. Yang, R. Wu, L. Shen, Y. P. Feng, Y. Dai, and B. Huang, Physical Review B **81,** 125211 (2010).
8. N. H. Hong, J. Sakai, N. Poirot, and V. Brizé, Physical Review B - Condensed Matter and Materials Physics **73,** 1-4 (2006).
9. W. A. Adeagbo and et al., Journal of Physics: Condensed Matter **22,** 436002 (2010).

10. Q. Xu, H. Schmidt, S. Zhou, K. Potzger, M. Helm, H. Hochmuth, M. Lorenz, A. Setzer, P. Esquinazi, C. Meinecke, and M. Grundmann, Applied Physics Letters **92**, 082508 (2008).

11. C. Martínez-Boubeta, J. I. Beltrán, L. Balcells, Z. Konstantinović, S. Valencia, D. Schmitz, J. Arbiol, S. Estrade, J. Cornil, and B. Martínez, Physical Review B - Condensed Matter and Materials Physics **82** (2010).

12. C. M. Araujo, M. Kapilashrami, X. Jun, O. D. Jayakumar, S. Nagar, Y. Wu, C. Arhammar, B. Johansson, L. Belova, R. Ahuja, G. A. Gehring, and K. V. Rao, Applied Physics Letters **96**, 232505-3 (2010).

13. T. Dietl, A. Haury, and Y. M. D'Aubigné, Physical Review B - Condensed Matter and Materials Physics **55** (1997).

14. X. H. Xu, H. J. Blythe, M. Ziese, A. J. Behan, J. R. Neal, A. Mokhtari, R. M. Ibrahim, A. M. Fox, and G. A. Gehring, New Journal of Physics **8**, 135 (2006).

15. S. J. Pearton, D. P. Norton, M. P. Ivill, A. F. Hebard, J. M. Zavada, W. M. Chen, and I. A. Buyanova, IEEE Transactions on Electron Devices **54**, 1040-1048 (2007).

16. M. Singh, H. M. Haverinen, P. Dhagat, and G. E. Jabbour, Advanced Materials **22**, 673-685 (2010).

17. Y. Wu, K. V. Rao, W. Voit, T. Tamaki, O. D. Jayakumar, L. Belova, Y. S. Liu, P. A. Glans, C. L. Chang, and J. H. Guo, IEEE Transactions on Magnetics **46**, 2152-2155 (2010).

18. Y. Wu, Y. Zhan, M. Fahlman, M. Fang, K. V. Rao, and L. Belova, in *'In-situ' solution processed room temperature ferromagnetic MgO thin films printed by inkjet technique*, (Mater. Res. Soc. Proc. **1292**, Boston, MA, 2010,) pp. 105-109.

19. R. Shannon, Acta Crystallographica Section A **32**, 751-767 (1976).

20. M. Kapilashrami, J. Xu, V. Strom, K. V. Rao, and L. Belova, Applied Physics Letters **95**, 033104-3 (2009).

21. B.B Straumal, A. A. Mazilkin, S. G. Protasova, A.A. Myatiev, P.B. Straumal, Gisela Schütz, P.A. Aken, E. Goering and B. Baretzky, Physical Review B **79**, 205206 (2009).

Mater. Res. Soc. Symp. Proc. Vol. 1394 © 2011 Materials Research Society
DOI: 10.1557/opl.2011.1532

# Experimental Evidence for Nitrogen as a Deep Acceptor in ZnO

M.C. Tarun,[1] M. Zafar Iqbal,[2] M.D. McCluskey,[1] J. Huso,[3] and L. Bergman[3]
[1]Department of Physics and Materials Science Program, Washington State University
Pullman, WA 99164-2814, U.S.A.
[2] Department of Physics, COMSATS Institute of Information Technology
Islamabad 44000, Pakistan
[3] Department of Physics, University of Idaho
Moscow, ID 83844, U.S.A.

## ABSTRACT

While zinc oxide is a promising material for blue and UV solid-state lighting devices, the lack of $p$-type doping has prevented ZnO from becoming a dominant material for optoelectronic applications. Over the past decade, numerous reports have claimed that nitrogen is a viable $p$-type dopant in ZnO. However, recent calculations by Lyons, Janotti, and Van de Walle [Appl. Phys. Lett. **95**, 252105 (2009)] suggest that nitrogen is a *deep* acceptor. In our work, we performed photoluminescence (PL) measurements on bulk, single crystal ZnO grown by chemical vapor transport. Nitrogen doping was achieved by growing in ammonia. In prior work at room temperature, we observed a broad PL band at ~1.7 eV, with an excitation threshold of ~2.2 eV, consistent with the calculated configuration-coordinate diagram. In the present work, at liquid-helium temperatures, the PL emission increases in intensity and red-shifts by ~0.2 eV. A peak is observed at 3.267 eV, which we tentatively attribute to an exciton bound to a nitrogen acceptor. Our experimental results indicate that nitrogen is indeed a deep acceptor and cannot be used to produce $p$-type ZnO.

## INTRODUCTION

Zinc oxide (ZnO) is a direct-gap semiconductor with a wide band gap (~3.4 eV at 300 K) and efficient excitonic emission at room temperature [1]. It is a promising material for various electronic and optical applications, including blue/UV light-emitting diodes and lasers. A reliable $p$-type dopant is, however, an outstanding problem that must be overcome for practical device applications [2]. Nitrogen substituting on a host oxygen site, $N_O$, has been discussed as a possible acceptor dopant (Table I). Many experimental [3,4,5] investigations suggested that $N_O$ has a shallow acceptor level in ZnO, with an acceptor binding energy of ~200 meV. Theoretical studies suggested a somewhat deeper level, roughly 0.3 to 0.5 eV above the valence-band maximum [6,7].

As reviewed in Ref. 2, reports of $p$-type ZnO are controversial. Like sightings of UFOs or Elvis, reports appear on a weekly or monthly basis. Table I lists several highly cited experimental papers that have reported $p$-type ZnO via doping with group-V elements. While these results initially seemed promising, repeatability has been a problem. It has been pointed out that the Hall effect, for example, can give the "wrong" carrier type if a sample is inhomogeneous [8].

To address these issues, Lyons, Janotti, and Van de Walle [9] calculated the properties of nitrogen in ZnO using density functional theory (DFT) with hybrid functionals. Their calculations show that $N_O$ is a deep acceptor, with the (0/-) acceptor level 1.3 eV above the valence band maximum. Optical absorption and emission energies of 2.4 eV and 1.7 eV, respectively, were obtained from the calculated configuration-coordinate diagram. The Stokes shift of 2.4 - 1.7 = 0.7 eV is due to large lattice relaxation of the deep acceptor. Lany and Zunger [10] have also obtained a deep level for $N_O$ using generalized Koopmans DFT. These theoretical results suggest that the reports of $p$-type conductivity, achieved with nitrogen doping, need to be re-evaluated.

Table I. Papers reporting $p$-type ZnO via group-V acceptor doping. Number of citations is from ISI Web of Science, November 1, 2011

| Paper | Citations |
|---|---|
| Look *et al.*, "Characterization of homoepitaxial p-type ZnO grown by molecular beam epitaxy," *Appl. Phys. Lett.* **81**, 1830 (2002). | 797 |
| Kim *et al.*, "Realization of p-type ZnO thin films via phosphorus doping and thermal activation of the dopant," *Appl. Phys. Lett.* **83**, 63 (2003). | 407 |
| Minegishi *et al.*, "Growth of p-type zinc oxide films by chemical vapor deposition," *Jpn. J. Appl. Phys. Part 2* **11A**, L1453 (1997). | 400 |
| Ryu *et al.*, "Synthesis of p-type ZnO films," *J. Crystal Growth* **216**, 330 (2000). | 360 |
| Ryu *et al.*, "Properties of arsenic-doped p-type ZnO grown by hybrid beam deposition," *Appl. Phys. Lett.* **83**, 87 (2003). | 311 |

**EXPERIMENT**

Bulk, single crystal ZnO samples were grown by chemical vapor transport in an ammonia ($NH_3$) ambient. The ammonia introduced $N_O$ acceptors and $N_O$–H pairs. A reference sample was grown in argon. Sub-gap photoluminescence (PL) and photoluminescence excitation (PLE) spectra were obtained using a JY-Horiba FluoroLog-3 spectrofluorometer with a 450-W xenon lamp as the excitation source. Low temperature measurements were performed using a Janis closed-cycle helium cryostat. Near-band-edge PL spectra were obtained with a He-Cd Kimmon laser (wavelength 325 nm) and a JY-Horiba micro-PL system consisting of a high-resolution triple monochromator and a UV microscope that can focus to a 1-$\mu$m diameter spot size. The low temperature measurements, using this setup, were carried out at 80 K in an INSTEC HCS621V UV-compatible microcell.

**RESULTS**

Previous measurements performed at room temperature are described in Ref. 11. ZnO:N samples exhibit a broad "red" PL emission band, peaking near 1.7 eV, with an excitation onset of

~2.2 eV, in good agreement with theoretical calculations. When samples are annealed, the intensity of the red emission band increases, as the $N_O$–H pairs are broken up and the $N_O$ acceptors are activated. The deep-acceptor behavior of nitrogen can be explained by the low position of the ZnO valence band relative to vacuum. Because of the low valence band, the $N_O$ acceptor level does not lie in the valence band, but rather 1.3-1.4 eV above the valence-band maximum.

In the present work, to verify that red luminescence is actually due to nitrogen, we compared spectra from samples grown in ammonia and argon. These are referred to as N-doped and undoped ZnO, respectively. For the N-doped sample, the N-H peak decreases as the sample is annealed in oxygen (Fig. 1). This is accompanied by a corresponding increase in the red luminescence. Spectra for the undoped sample were multiplied by 10 for clarity. From Fig. 1, one can see that the N-H peak and red luminescence are both an order of magnitude weaker than for the deliberately N-doped sample. (Unintentional N doping probably came from contamination during the growth). This strong effect provides further evidence that the red luminescence is definitely correlated with nitrogen acceptors.

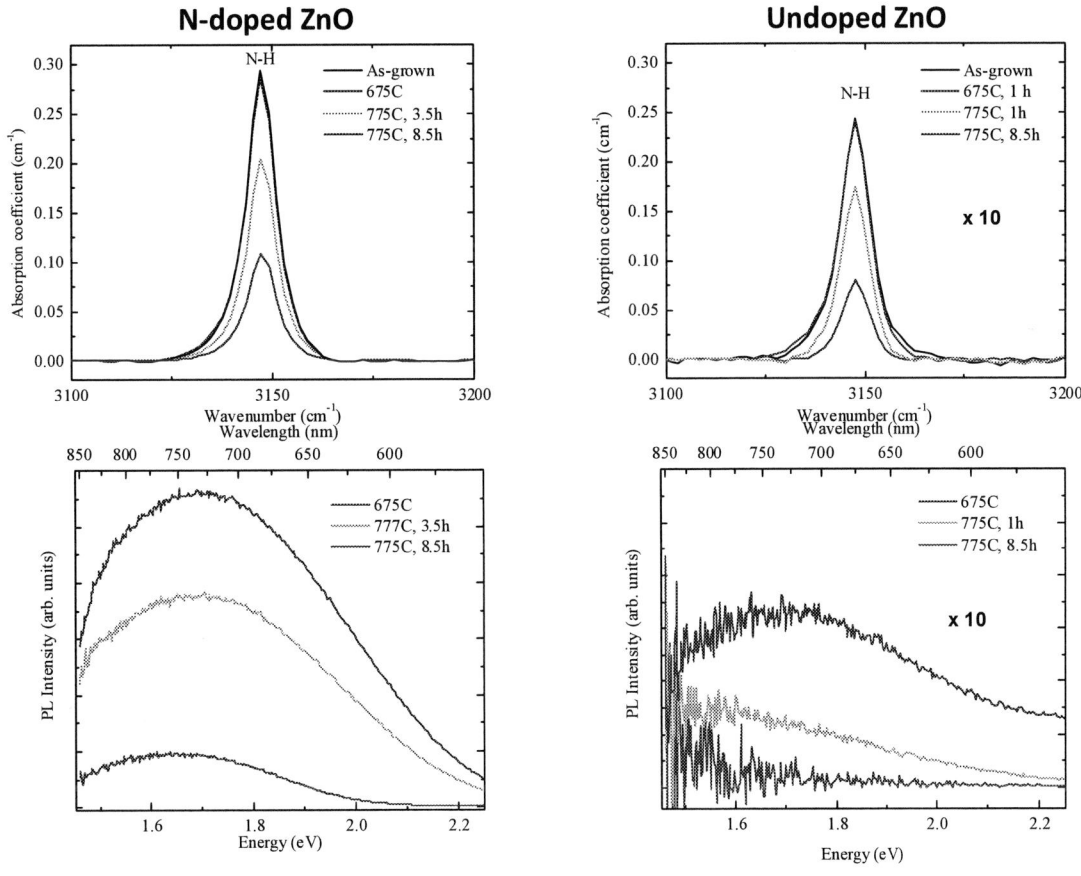

Figure 1. Room-temperature spectra for N-doped and nominally undoped ZnO. Note that the undoped ZnO spectra have been multiplied by 10. Top: IR absorption, showing a peak due to N-H pairs. Bottom: Red luminescence, attributed to N acceptors.

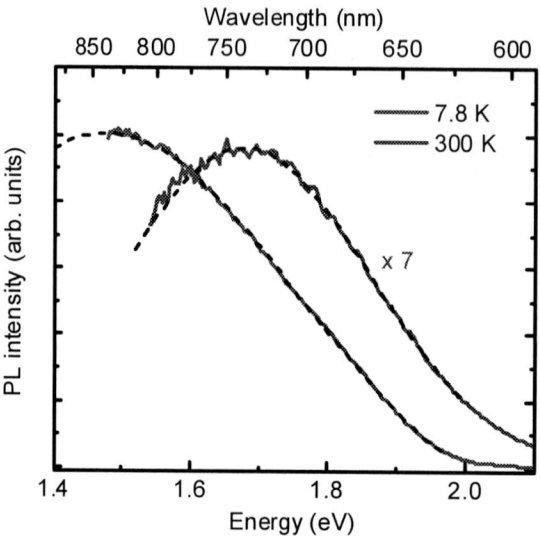

Figure 2. PL spectra of N-doped ZnO.

The red luminescence increases in intensity as the temperature is lowered. As shown in Fig. 2, the peak also exhibits a red-shift. The center of the peak shifts from the red (1.7 eV) to the near-IR (1.5 eV) as the temperature decreases from 300 to 7.8 K. Several factors may contribute to this shift. First, as the temperature is lowered, one goes from a band-to-acceptor to a donor-acceptor transition, the latter having a lower energy. Second, the thermal change in population of vibrational states may affect the PL energies.

The near-band-edge PL spectra are shown in Fig. 3. For the N-doped sample (annealed at 775°C), the free exciton (3.378 eV), donor bound exciton (3.363), and phonon replicas (1LO, 2LO) are clearly resolved. These peaks are also seen in the undoped sample, although they are broader. The spectrum for the N-doped sample contains a peak at 3.267 eV, which is absent in the undoped sample. This peak has phonon replicas near 3.194 and 3.124 eV. We tentatively attribute this new peak to a nitrogen acceptor-bound exciton. Given this assignment, the exciton binding energy is 0.1 eV. This value is ~1/10 that of the acceptor ionization energy, which follows Haynes' rule [12] for exciton binding energies in semiconductors. A PL peak with similar zero-phonon energy, 3.235 eV, was previously attributed to nitrogen acceptors [4]. Further work will be required to provide a definitive assignment.

The PL intensity increased and the peak linewidths decreased with annealing, indicating improved quality. Specifically, the 3.267 eV peak was not observed in the as-grown ZnO:N sample but is observed in the sample annealed at 775°C. This correlation, similar to that found for the red luminescence, is consistent with the dissociation of N-H pairs.

**CONCLUSIONS**

In conclusion, by comparing the PL spectra for the undoped and N-doped ZnO samples, we have confirmed that the red luminescence is correlated with nitrogen acceptors. A PL peak at

3.267 eV (T = 80 K) is attributed to the N acceptor bound exciton. These results are in agreement with the deep N-acceptor model of Lyons, Janotti, and Van de Walle [9]. These experimental and theoretical results would appear to rule out substitutional nitrogen as a *p*-type (shallow level) dopant in ZnO. The other group-V dopants should lead to even deeper acceptor states. Possible avenues for *p*-type ZnO may involve tuning the valence band or using zinc vacancy– related acceptors [13].

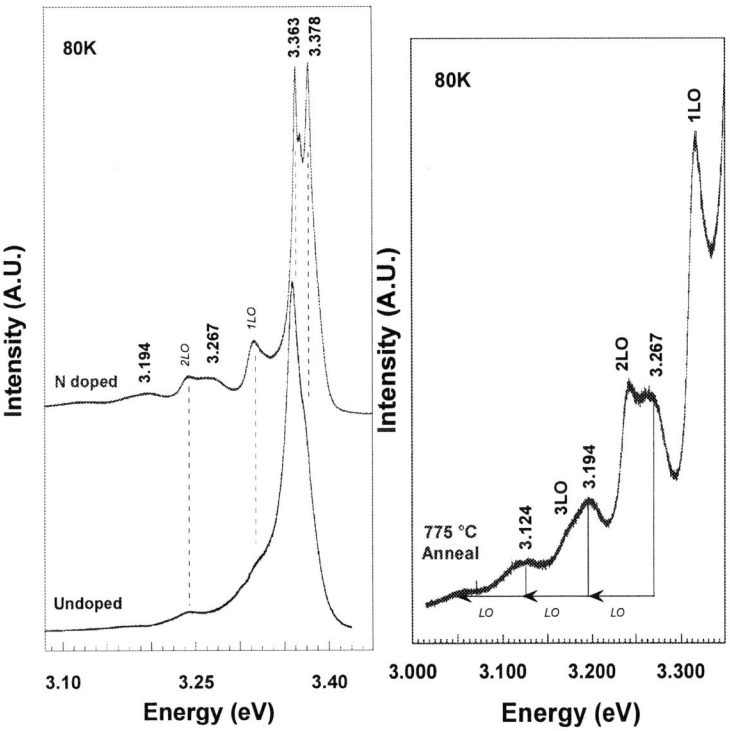

Figure 3. PL spectra near the ZnO band edge.

## ACKNOWLEDGMENTS

Funding for this work was provided by the National Science Foundation Grant No. DMR-1004804 (WSU) and Department of Energy Grant No. DE-FG02-07ER46386 (UI).

# REFERENCES

1. Y. Chen, D.M. Bagnall, H.-J. Koh, K.-T. Park, K. Hiraga, Z.-Q. Zhu, and T. Yao, J. Appl. Phys. **84**, 3912 (1998).

2. M.D. McCluskey and S.J. Jokela, J. Appl. Phys. **106**, 071101 (2009).

3. K. Thonke, T. Gruber, N. Teofilov, R. Schnfelder, A. Waag, and R. Sauer, Physica B **308–310**, 945 (2001).

4. A. Zeuner, H. Alves, D.M. Hoffman, B.K. Meyer, A. Hoffmann, U. Haboeck, M. Strassburg and M. Dworzak, Phys. Status Solidi B **234**, R7 (2002).

5. L. Wang and N.C. Giles, Appl. Phys. Lett. **84**, 3049 (2004).

6. X.M. Duan, C. Stampfl, M.M.M. Bilek, D.R. McKenzie, and S.-H. Wei, Phys. Rev. B **83**, 085202.

7. C.H. Park, S.B. Zhang, and S.-H. Wei, Phys. Rev. B **66**, 073202 (2002).

8. O. Bierwagen, T. Ive, C.G. Van de Walle, and J.S. Speck, Appl. Phys. Lett. **93**, 242108 (2008).

9. J.L. Lyons, A. Janotti, and C.G. Van de Walle, Appl. Phys. Lett. **95**, 252105 (2009).

10. S. Lany and A. Zunger, Phys. Rev. B **81**, 205209 (2010).

11. M.C. Tarun, M. Zafar Iqbal, and M.D. McCluskey, AIP Advances **1**, 022105 (2011).

12. J.R. Haynes, Phys. Rev. Lett. **4**, 361 (1960).

13. S.T. Teklemichael, W.M. Hlaing Oo, M.D. McCluskey, E.D. Walter, and D.W. Hoyt, Appl. Phys. Lett. **98**, 232112 (2011).

Mater. Res. Soc. Symp. Proc. Vol. 1394 © 2012 Materials Research Society
DOI: 10.1557/opl.2012.246

# Nitrogen pair − hydrogen complexes in ZnO and p-type doping.

Adisak Boonchun,[1] Walter R. L. Lambrecht,[1] Jiraroj T-Thienprasert,[2,3] and Sukit Limpijumnong[3,4]
[1]Department of Physics, Case Western Reserve University, Cleveland, OH 44106-7079, USA
[2]Department of Physics, Kasetsart University, Bangkok 10900, Thailand
[3]Thailand Center of Excellence in Physics (ThEP Center), Commission on Higher Education, Bangkok 10400, Thailand
[4]School of Physics, Suranaree University of Technology and Synchrotron Light Research Institute, Nakhon Ratchasima 30000, Thailand

## ABSTRACT

Electronic structure calculations using the Heyd-Scuseria-Ernzerhof (HSE) hybrid functional are presented for $N_O$-pair complexes with and without hydrogen to test the hypothesis that such defect complexes could lead to shallower levels than for isolated $N_O$ and hence p-type doping. The H is found to bind strongly to one of the N in the pair and removes the corresponding defect level from the gap but the second N's polaronic defect level in the gap remains deep.

## INTRODUCTION

The main bottleneck for the full deployment of ZnO as an optoelectronic material remains the lack of reliable p-type doping. Although several alternative doping strategies have been proposed, nitrogen doping on the oxygen site remains of interest. Although, based on *ab-initio* calculations [1,2], isolated $N_O$ is now well accepted to have a very deep energy level and hence to be unsuitable for p-type doping, it was recently pointed out by Meyer et al. [3] that in N-doped samples a shallow acceptor level exists at about 170 meV above the valence band maximum (VBM). Recently, Lautenschlaeger et al. [4] proposed that with high enough N concentration, one could generate a sufficient number of pairs, which could be stabilized by hydrogen and proposed this complex as a potential candidate for this shallow acceptor level. Evidence for donor-acceptor pair interaction, which is known to potentially make acceptor levels defect shallower was pointed out in photoluminescence spectra in their paper. In this paper we investigate this hypothesis with first-principles calculations. We investigate the optimum location for the hydrogen in the complex, the stability of the complex, and whether it makes the levels shallower.

## COMPUTATIONAL METHOD

Our study is based on density functional calculations using the Heyd-Scuseria-Ernzerhof (HSE) hybrid functional [5]. The calculations are performed using the VASP code [6] and using the projector augmented wave (PAW) method [7]. Supercells of 72, 96 and 128 atoms are used to monitor the convergence of the defect levels as function of supercell size. We also investigate the convergence with k-point sampling as detailed below in the results, and use a well converged energy cut-off of 300 eV for the projector augmented plane waves.

# RESULTS

We start our investigation with a study of the single substitutional $N_O$ as function of supercell size (number of atoms per cell), k-point sampling (Monkhorst-Pack [8] mesh n×n×n and Hartree-Fock mixing parameter $\alpha$ of the HSE functional. Table 1 lists our results. Note that $\alpha$ = 0.375 leads to a Γ-point band gap of 3.38 eV, close to the experimental value. We conclude that a 96 atom cell is well converged in terms of size and the 1×1×1 sampling is sufficient for this size cell. We also note that our results for the 72 atom cell and $\alpha$ = 0.36 are close to those of Lyons et al. [1] (~1.3 eV). However, with a larger cell and $\alpha$-value that better reproduces the gap of ZnO, we obtain an even deeper transition level than before. Because the level is deep, we do not include a defect band dispersion correction.

Table 1: Transition level for single $N_O$ calculated with different computational parameters.

| Cell size | k-point sampling | $\alpha$ | E 0/− (eV) |
|---|---|---|---|
| 72 | 222 | 0.375 | 1.54 |
| 72 | 222 | 0.360 | 1.47 |
| 96 | 111 | 0.360 | 1.75 |
| 96 | 111 | 0.375 | 1.82 |
| 96 | 222 | 0.375 | 1.81 |
| 128 | 111 | 0.360 | 1.73 |

As is well known, this extremely deep level compared to the generalized gradient approximation (GGA) result of about 0.4 eV, results from the localization of the hole in the neutral charge state on a single N-p orbital, the $p_z$ orbital, which has $a_1$ symmetry for the $C_{3v}$ symmetry defect [2]. It is accompanied by a polaronic distortion, in which one N-Zn bond length is extended compared to the other ones ($d_\parallel / d_\perp \approx 1.06$). HSE allows this symmetry breaking solution by having an orbital dependent potential, which lowers occupied states versus unoccupied ones. The one-electron levels show some perturbation of the VBM and a minority-spin level quite high in the gap at energy larger than 2 eV.

Next, we consider a $N_O$-pair without H. We find that this defect has two transition levels, a 0/− at 1.68 eV and a −/2− one at 2.13 eV, both of which are still very deep in the gap. In contrast to the single $N_O$, it has no net magnetic moment, meaning that the electrons have opposite spin in the neutral charge state. This leads to a one-electron level located somewhat lower in the gap. These results indicate that the polaronic distortion is maintained for both and no significant covalent bonding occurs between the two defect states. In fact, in a simple molecular model, one would expect the N-defect orbitals to couple mostly through the common Zn in between them. In a simple triatomic molecule of this type one expects, by perturbation theory, that if the phases of the two N-orbitals are the same, they would lead to a shallower level. However, when the phases are opposite, they cancel each other and the level stays the same. So, the available empty level for accepting electrons stays as deep as in the isolated $N_O$. As far as total energy is concerned, we find the pair in the neutral charge state to be slightly stable with respect to the isolated neutral $N_O$ but in the double negative charge state it is unstable by 0.18 eV with respect to the corresponding negative isolated $N_O$. As expected, negatively charged defects repel each other.

Next, we consider a single $N_O$ with a hydrogen. This system was previously studied in the local density approximation (LDA) by Limpijumnong et al. [9]. They found the H prefers the antibond position $AB_{\parallel}$. We find similar results in HSE. The main effect of the H is that it passivates the N-defect orbital and pushes it all the way in the valence band resulting in a gap free of defect levels. The H bonds strongly to the N within a bond length of about 1 Å and the donor acceptor pair interaction is so strong that the level does not become shallow but disappears altogether.

For the N-pair with hydrogen, which we label $N_O^2$-H, we consider various positions for the H and N as shown in Fig. 1.

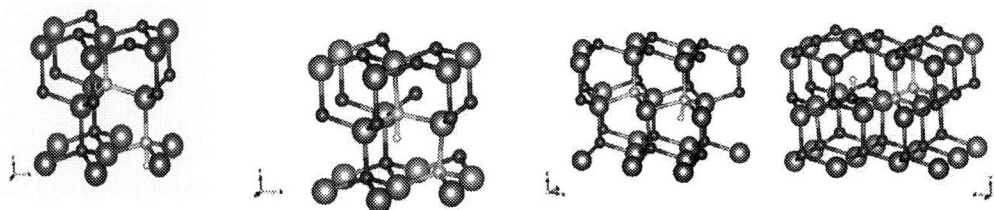

Figure 1: Models 1-4 from left to right for $N_O^2$-H: relaxed structures in neutral charge state. In models 1 and 2, the two N lie in different basal planes with H as indicated, in model 3 the two N lie in the same basal plane. In model 4, the H is farther away from the $N_O^2$ complex. Grey large sphere: Zn, red small sphere: O, small blue sphere : N, smallest pink sphere: H.

Figure 2 shows the total density of states for model 1 compared to that of bulk ZnO in both charge states. We can see a minority spin peak slightly below the conduction band minimum very similar to the results for a single $N_O$. The latter shows up as a single small peak because of our sparse Γ-point sampling: the energy levels are simply broadened by Gaussians and allow us to place the VBM and CBM precisely. Figure 3 shows the spin density corresponding to this state and shows the typical very localized hole of the polaronic state. The conclusion of this is that in the all $N_O^2$-H complexes studied, the H simply passivates one of the N in the pair and the other maintains its deep polaronic state. Similar results are obtained for the other models, i.e., models 2 and 3 (not shown here). The 0/− transition levels are found at 1.82, 1.83, 1.80, 2.14 eV for models 1-4 respectively.

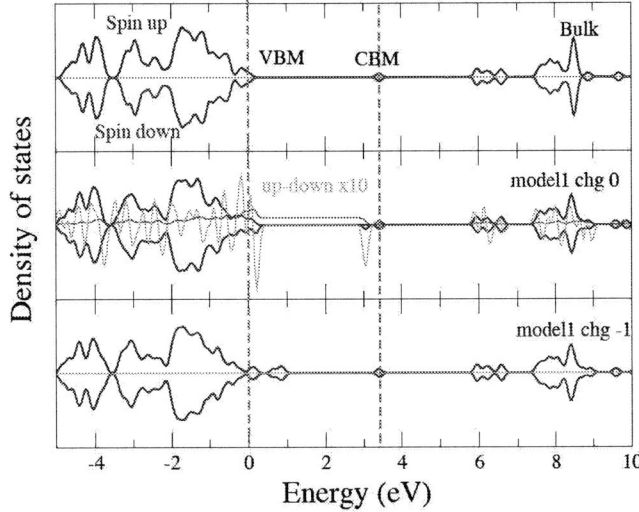

Figure 2: Density of states for N-pair model in neutral and negative charge states compared with bulk ZnO. In the central panel, the green line shows the difference between up (blue) and down (red) spin multiplied by a factor 10 for better visibility. The violet line is its cumulative integral showing that there is a net moment of 1 $\mu_B$.

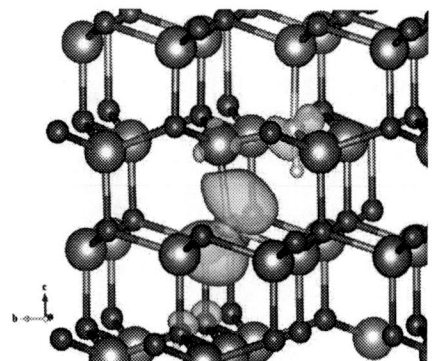

Figure 3: Spin density for the minority spin gap state in the N-pair model 1. The wave function is seen to be localized on the N which is not bonded to the H. The H is indicated by the small pink sphere, the Zn atoms by the large gray spheres, the O by the small red spheres and the N by the small blue spheres. The spin density isosurface is shown by the yellow surface.

To investigate whether a weaker donor acceptor pair interaction of the hydrogen to the N-pair could lead to shallower levels we also investigated the model 4 shown in Figure 4. Although we find somewhat different states, as explained in the figure caption, we still find a very deep level acceptor level for the minority spin state. It has more or less the anticipated molecular orbital character discussed earlier for the N-pair.

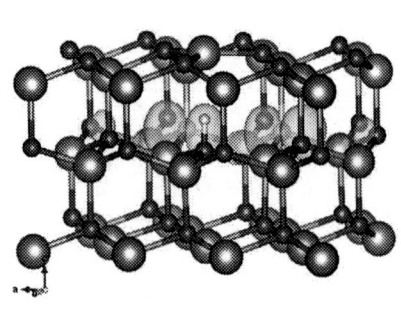 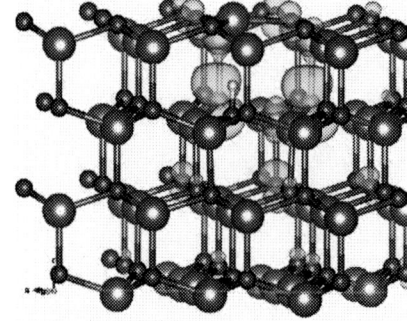

Figure 4: minority (left) and majority (right) spin densities for model 4 with H in Zn-O bond center. The most localized state in this case is again for minority spin but consist of in-plane N-orbitals shared with the Zn in between.

Returning to the total energy question, we find that the neutral $N_O^2$-H complex is significantly more stable (by about 0.29 eV) than the separate neutral $N_O$ plus a separate $H_{BC}^+$ (hydrogen in the bond center in the positive charge state which is the stable form of H at any Fermi level) plus an electron at the Fermi level, as described by its binding energy,

$$E_b(complex) = E_f(N_O^2H) - 2E_f(N_O) - E_f(H_{BC}^+) - \mu_e$$

(1)

This indicates that indeed the H helps to stabilize the N-pairs as anticipated by Lautenschlaeger et al. [4].

**DISCUSSION**

Lautenschlaeger et al.'s model [4] is at first sight quite appealing because it could potentially explain a number of aspects of N-doping. If their model were correct, it would predict an extreme sensibility to H concentration. Too much H would lead to compensation and n-type materials, too little H would lead to too few pairs forming and hence no p-type. Secondly, one would expect that at the high concentration of N of about $10^{20}$ cm$^{-3}$ required to have of the order of $10^{18}$ cm$^{-3}$ pairs, the VBM would be significantly perturbed by an impurity band tail resulting from the N-orbitals on which the hole is not localized. One can see these states in the density of states as shown in Figure 1 above the VBM. One would expect that if hole conduction occurs in such a band tail, it would exhibit a very low mobility. This would be consistent with experimental data on mobility in the few cases where p-type doping was achieved [10]. One might also expect that the H could be unstably bound in the complex and thereby account for loss of p-type conductivity over time.

Our calculations indicate that the N-pair can indeed form a stable pair with H. However, we find no indications of this leading to a shallow level. The H simply passivates one of the N leaving the other in its deep level polaronic state. Even with somewhat more distant H in a nearby Zn-O bond center, we find essentially a similar electronic structure with deep polaronic states as expected for a N-pair without H. It appears that other complexes involving N need to be identified to explain the observed shallow level. More tightly bound N-pairs as in a $N_2$ molecule have been investigated by Limpijumnong et al. [11]. However, when such a pair is put on an O-site as in a split interstitial configuration, it leads to a double donor rather than an acceptor.

## CONCLUSIONS

Calculations using the HSE hybrid functional in larger supercells than used before confirmed a deep level for $N_O$, in fact an even deeper level than before. N-pairs would not lead to a shallower level due to covalent bonding between their defect orbitals because the latter bind only via the Zn in between. Hydrogen helps stabilize the complex but passivates one of the N-defect orbitals by forming a strong bond with it in the antibonding site and leaves the other N in its deep polaronic state. Therefore, the acceptor level of the complex is not shallower than the isolated N and cannot lead to p-type doping

## ACKNOWLEDGMENTS

We wish to thank Bruno K. Meyer for telling us about their model. J. T. thanks the B. K. Meyer group for hospitality. The work at CWRU was supported by NSF under grant number DMR-1104595.

## REFERENCES

1. J. L. Lyons, A. Janotti, and C. G. Van de Walle, *Appl. Phys. Lett.* **95**, 252105 (2009).
2. S. Lany and A. Zunger, *Phys. Rev. B* **81**, 205209 (2010).
3. B. K. Meyer, J. Sann, D. M. Hofmann, C. Neumann and A. Zeuner, *Semicond. Sci. Technol.* **20**, S62 (2005).
4. S. Lautenschlaeger, M. Hofmann, S. Eisermann, G. Haas, M. Pinnisch, A. Laufer, and B. K. Meyer, *Phys. Stat. Solidi B* **248**, 1217 (2011).
5. J. Heyd, G. E. Scuseria, and M. Ernzerhof, *J. Chem. Phys.* **124**, 219906 (2006).
6. G. Kresse and J. Furthmüller, *Comput. Mat. Sci.* **6**, 15-50 (1996).
7. P. E. Blöchl, *Phys. Rev. B* **50**, 17953 (1994).
8. H. J. Monkhorst and J. D. Pack, *Phys. Rev. B* **13**, 5188 (1976)
9. S. Limpijumnong, X. Li, S.-H. Wei, and S. B. Zhang, *Physica B* **376-377**, 686 (2006).
10. Z.-Q. Fang, B. Claflin, D. C. Look, L. L. Kerr, and X. Li, *J. Appl. Phys.* **102**, 023714 (2007).
11. S. Limpijumnong, X. Li, S.-H. Wei, and S. B. Zhang, *Appl. Phys. Lett.* **86**, 211910 (2005).

Mater. Res. Soc. Symp. Proc. Vol. 1394 © 2012 Materials Research Society
DOI: 10.1557/opl.2012.257

# Growth Techniques for Bulk ZnO and Related Compounds

Detlef Klimm, Detlev Schulz, Steffen Ganschow, Zbigniew Galazka, Reinhard Uecker

Leibniz Institute for Crystal Growth, Max-Born-Str. 2, 12489 Berlin, Germany

## ABSTRACT

ZnO bulk crystals can be grown by several methods. 1) From the gas phase, usually by chemical vapor transport. Such CVT crystals may have high chemical purity, as the growth is performed without contact to foreign material. The crystallographic quality is often very high (free growth). 2) From melt fluxes such as alkaline hydroxides or other oxides ($MoO_3$, $V_2O_5$, $P_2O_5$, PbO) and salts ($PbCl_2$, $PbF_2$). Melt fluxes offer the possibility to grow bulk ZnO under mild conditions (<1000°C, atmospheric pressure), but the crystals always contain traces of solvent. The limited purity is a severe drawback, especially for electronic applications. 3) From hydrothermal fluxes, usually alkaline (KOH, LiOH) aqueous solutions beyond the critical point. Due to the amphoteric character of ZnO, the supercritical bases can dissolve it up to several per cent of mass. The technical requirements for this growth technology are generally hard, but this did not hinder its development as the basic technique for the growth of α-quartz, and meanwhile also of zinc oxide, during the last decades. 4) From pure melts, which is the preferred technology for numerous substances applied whenever possible, e.g. for the growth of silicon, gallium arsenide, sapphire, YAG. The benefits of melt growth are (i) the high growth rate and (ii) the absence of solvent related impurities. In the case of ZnO, however, it is difficult to find container materials that are compatible from the thermal (fusion point $T_f = 1975$°C) and chemical (required oxygen partial pressure) point of view. Either cold crucible (skull melting) or Bridgman (with reactive atmosphere) techniques were shown to overcome the problems that are inherent to melt growth. Reactive atmospheres allow to grow not only bulk ZnO single crystals, but also other TCOs such as β-$Ga_2O_3$ and $In_2O_3$.

## INTRODUCTION

The semiconducting properties of zinc oxide have been used for years for the production of voltage dependent resistors (varistors). In such devices *p-n* junctions are formed in the grain boundaries of ceramic bodies. Moreover, doped ZnO can be used as TCO (transparent conducting oxide) e.g. for transparent electrodes. Still the production of single crystalline heteropolar devices, especially LED's, is prohibited by the fact that efficient *p*-type doping (high mobility and sufficient hole density) was not yet possible. Further work towards ZnO based devices requires not only improved epitaxy techniques, but also the delivery of substrates for homoepitaxy with good quality.

Principally, ZnO single crystals and substrates made thereof are available nowadays. The huge variety of growth techniques that can be used for the production of bulk ZnO is surprising, and perhaps no other substance is really grown by so many different methods. The benefits and drawbacks of these methods will be reviewed in this paper, and a short outlook on the bulk growth of alternative TCO materials will be given.

## GROWTH TECHNIQUES FOR ZnO SORTED BY PHASE TRANSITION

The chemical substance zinc oxide is a white powder, which is only apparently chemically stable. Instead, it decomposes easily to its components (1), or reacts with other substances such as carbon dioxide (2) or water (3):

$$Zn + \frac{1}{2}O_2 \leftrightarrow ZnO \quad \Delta H = -350.5\,kJ/mol \tag{1}$$

$$ZnO + CO_2 \leftrightarrow ZnCO_3 \quad \Delta H = -68.6\,kJ/mol \tag{2}$$

$$ZnO + H_2O \leftrightarrow Zn(OH)_2 \quad \Delta H = +2.4\,kJ/mol \tag{3}$$

(Enthalpies are given for normal conditions). Carbon dioxide and water vapor are with 390 ppm or ≈10,000 ppm, respectively, constituents of ambient air — hence the surface of ZnO powder as well as ZnO crystals is usually covered by layers of zinc carbonate and/or hydroxide (Fig. 1).

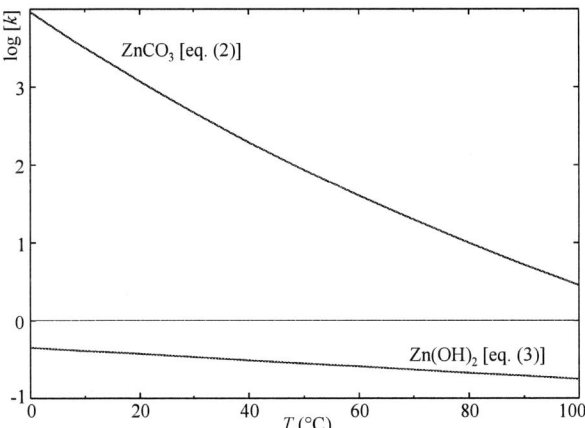

**Fig. 1: Equilibrium constants $k(T)$ for the chemical reactions (2) and (3). It is obvious that ZnO if stored under ambient air is always contaminated by carbonate and to a smaller degree by hydroxide.**

Zinc hydroxide (3) is amphoteric. This means it can easily be dissolved by weak aqueous acids (4) and by strong bases (5):

$$Zn(OH)_2(aq) + 2\,H^+(aq) \leftrightarrow Zn^{2+}(aq) + 2\,H_2O \quad \Delta H = -203.8\,kJ/mol \tag{4}$$

$$Zn(OH)_2(aq) + OH^-(aq) \leftrightarrow [Zn(OH)O]^-(aq) + H_2O \quad \Delta H = +8.7\,kJ/mol \tag{5}$$

Not only hydroxide zincates such as $[Zn(OH)_4]^{2-}$ or $[Zn(OH)O]^-$ (5) can be formed in solution, but also other complex ions such as ammono zincates e.g. $[Zn(NH_3)_4]^{2+}$.

The high reactivity of zinc metal and zinc oxide, which is expressed by the equilibrium reactions (1) – (5), allows the application of a big variety of growth techniques for ZnO bulk crystals. Nearly all technical processes that were developed for growing bulk single crystals can be applied for this material. In the next subsections, these techniques for ZnO bulk crystal growth will be categorized according to the phase transformation that enables the crystallization of ZnO from a fluid mother phase.

Besides it should be mentioned that diethylzinc $(C_2H_5)_2Zn$ is the first organometallic compound that was ever prepared (in 1849 by Frankland from Zn metal and ethyl iodide [1]). Compounds of this type are typically used for the MOCVD (metal organic chemical vapor deposition) of thin layers, but were so far not used for bulk crystal growth.

### **From the gas phase**

The formation of ZnO from Zn metal and oxygen is strongly exothermal (1): Liquid zinc burns in air under the formation of white ZnO fume. On an industrial scale, Zn metal can be evaporated and reacts with air in the indirect (French) process to ZnO. Upon heating, ZnO dissociates completely and the vapor pressure of its constituents Zn and $O_2$ becomes so large, that the solid starts to evaporate. The evaporation rate depends on the surrounding atmosphere: in thermogravimetry measurements the mass loss becomes remarkable at >1200°C in vacuum $(10^{-6}$ bar), >1300°C in argon, and >1450°C in air [2, Fig. 7 there]. For $T > 1600$°C the vapor pressure becomes so large that reasonable transport rates for sublimation growth (physical vapor transport, PVT) can be obtained [3]. It should be noted, however, that such crystals are typically only on the millimeter scale.

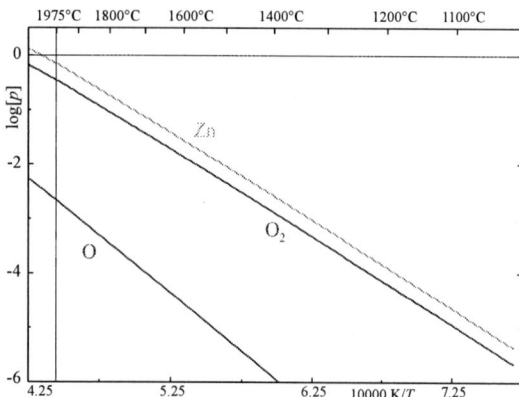

**Fig. 2: Vapor pressure of different species over ZnO. 1975°C is the melting point of ZnO. The horizontal line corresponds to 1 bar.**

Significantly better crystals can be grown by chemically assisted vapor phase transport (CVT). In Erlangen (Germany) the growth of bulk ZnO by different techniques was performed since the 1930s (see next subsection), and there E. Scharowsky [4] performed a PhD work based on an apparatus that is schematically drawn in Fig. 3. There liquid Zn metal (melting point 419°C) was held inside a ceramic tube at ca. 600°C. It should be noted that the volatility of Zn is way higher compared to ZnO, and reaches at 600°C already 15 mbar (boiling point at ambient pressure is ca. 1200°C). A flow of ca. 1100 ml/min oxygen-free nitrogen (with ca. 60 ml/min hydrogen added to avoid accidental Zn oxidation) transports the evaporated zinc via a conical tube outlet to a second, larger furnace that is held at higher temperature (maximum 1200°C). This furnace is partially open and allows the inflow of air. The $N_2$ flux must be adjusted in such a manner that backflow of air to the Zn metal is prohibited, but on the other side a too strong "blow" prohibits crystal growth at the rim of the cone. Under optimum conditions colorless ZnO needles with up to 4 cm length and a few tenth of a millimeter diameter with almost perfect hexagonal cross section could be obtained. Sometimes lancet shaped single crystals were harvested. A too large gas flow results in microcrystalline ZnO fume, under a too large $H_2$ flux the crystals become yellowish.

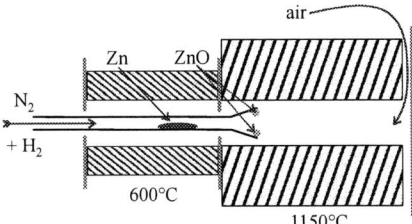

**Fig. 3: The setup that was used by Scharowsky (1953) [4] for the CVT growth of ZnO single crystals.**

Later this technique was modified by Helbig [5] who used annealed ZnO feed as source material. With similar $N_2/H_2$ flow rates through the feedstock, and a separately controlled flow of pure $O_2$ to the growth zone, ZnO crystals up to 20 g mass were grown (diameter up to 7 mm). One can assume that the considerably higher and well controlled growth temperature (1600±0.2 K instead of 1423 K) enabled the larger and more isotropic growth rate, compared to Scharowsky [4]. The material that was offered by Eagle-Picher, Inc. is produced in a similar process based on ZnO transport by $H_2$ [6].

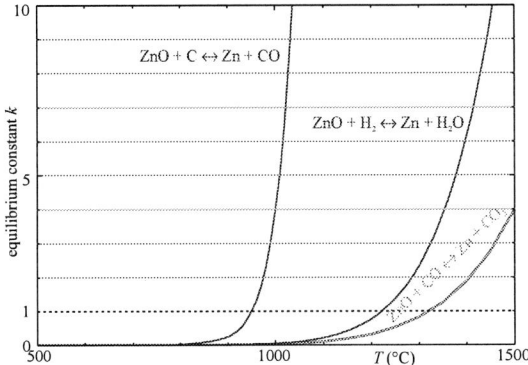

**Fig. 4: Equilibrium constants for different chemical reactions enabling CVT growth of ZnO (total pressure = 1 bar).**

Chemical vapor transport relies on an equilibrium reaction between the substance to be grown and some transport agent. If $H_2$ is used, like in the examples described so far, this is the reaction that labels the middle (blue) curve in Fig. 4. Reasonable growth rates require conditions yielding equilibrium constant $k \approx 1$. This is the case around 1200°C.

Admixtures of solid carbon to the feedstock (source temperature around 1000°C [7]) or carbon monoxide gas flow (source temperature up to 1300°C [8]) are alternative transport agents for CVT growth of ZnO and are depicted by the red and green curves in Fig. 4, respectively.

It is an advantage of all growth processes from the gas phase that the crystals grow mainly free from mechanical loads (except minor thermal stresses) and in a relatively clean environment where besides Zn and O only the transport agent is present. Consequently, the crystals often show superior quality by means of chemical purity as well as crystallographic quality (e.g. rocking curve width for 2Θ scans FWHM < 1 arcmin, carrier mobility >180 cm$^2$/Vs). Two aspects should be mentioned that might be less beneficial: In the case of carbon, C can be incorporated if the oxygen partial pressure in the system is too low [9], and hydrogen or hydrogen related point defects are known to play a dominant role in the electrical properties on ZnO [10].

## From the melt

Already 20 years before Scharowsky's [4] vapor growth experiments in the same laboratories in Erlangen annealing experiments were performed. Fritsch (1935) [11] pressed ZnO powder to cylinders with length up to 20 mm and diameter up to 9 mm (Fig. 5a). A resistance heater was used to heat up these cylinders to 800°C and reduce the electrical resistivity of the semiconducting material. In a second step, AC current up to 15 A was passed through the cylinders for several hours, resulting in a pyrometrically measured surface temperature up to 1500°C. These experiments were performed in different atmospheres under different total pressure, up to 120 bar of pure oxygen. Sometimes clear drops of molten and solidified ZnO appeared — obviously the first molten ZnO ever produced! Experiments to melt pure ZnO in an iridium container failed because Ir was oxidized. Similar experiments with *rf* heating were performed later by Burmeister [12].

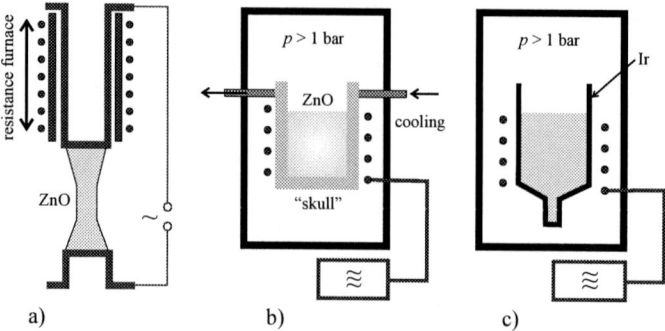

Fig. 5: 3 different methods that were developed for the growth of bulk ZnO from the melt: a) Direct heating of ZnO ceramic by AC current [11]. b) Skull melting by *rf* energy coupling directly into ZnO melt [14]. c) Bridgman growth with iridium crucible and reactive atmosphere [15].

Skull melting is an alternative method for melt growth which was developed for ZnO by Nause and is commercialized by CERMET Inc. [13,14]. Here *rf* heating energy is coupled directly into the ZnO melt that is surrounded by a cooled "skull" of solid ZnO powder (Fig. 5b). ZnO crystals grown by this method are chemically very pure because the material has no contact to any foreign substance. The atmosphere can be controlled in a very wide range, which is beneficial for the Zn/O stoichiometry of the grown crystals. Nevertheless, skull melting has one severe drawback: The electrical conductivity of dielectric melts rises usually with temperature. Consequently, "hot spots" inside the melt volume are heated more than colder volume elements, and thermal gradients are boosted.

Another concept for ZnO melt growth relies on "reactive atmospheres" with self-adjusting oxygen partial pressure $p_{O_2}(T)$ and reduces problems that can occur if iridium crucible metal is used in an oxidizing atmosphere [15,16]. A conflict arises from the dissociation reaction (1) that leads to a decomposition of ZnO at high $T$ which could be reduced (according the law of mass action) by a high $p_{O_2}$, and the oxidation reaction

$$Ir + O_2 \leftrightarrow IrO_2 \quad \Delta H = -240.2\,\text{kJ/mol} \tag{6}$$

where the remarkably negative $\Delta H$ shows that at room temperature iridium, although a member of the platinum group of metals, is thermodynamically unstable against oxygen.

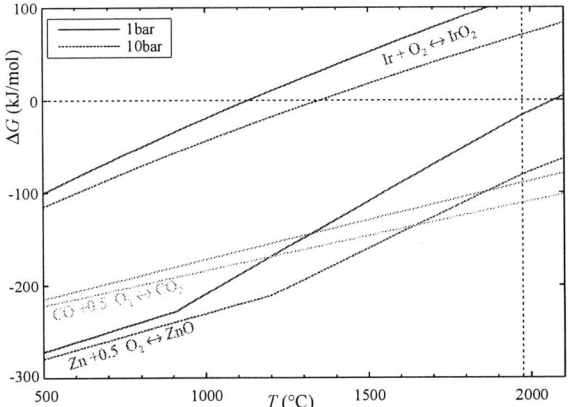

**Fig. 6: Gibbs free energy balance for the 3 reactions that are relevant for ZnO growth from Ir crucibles within reactive $CO_2$ atmosphere for 1 bar and 10 bar total pressure. The vertical dashed line marks the melting point of ZnO $T_f$ = 1975°C.**

The Gibbs free energy balance for ZnO (1) and $IrO_2$ (6) formation are plotted for two pressures in Fig. 6. It was already mentioned that the evaporation of ZnO under ambient pressure is so high that it sublimes without previous melting, but already under an overpressure of a few bar the ZnO melt can be kept stable. The following points are read from Fig. 6:

- Ir oxidation becomes unfavorable above ca. 1100-1300°C. Therefore for very high temperatures corrosion by oxygen will only be marginal.

- Zn oxidation is favored only for low $T$. The tendency of ZnO to dissociate to Zn + $O_2$ (see also Fig. 2) can be suppressed to a certain degree by application of overpressure.

- In the whole range where Ir could be oxidized ($T$ < 1300°C) the dissociation of ZnO is almost completely prohibited because $CO_2$ dissociates more. This means that in a $CO_2$ atmosphere no liquid Zn can be formed (boiling point ca. 1200°C) that would alloy and destroy the crucible.

For growing crystals, pressed ZnO powder is placed inside an Ir crucible with conical bottom and seed channel, as shown in Fig. 5c. The crucible is surrounded by ceramic insulation (not shown in the figure) and is heated by an *rf* induction coil. If the crucible content (except the seed on the bottom) is molten, the heating power is lowered to initiate seeded growth. So far, single crystalline ZnO boules up to 38 mm diameter were grown [17].

### From solutions, including hydrothermal

So far, three types of solvents were reported for ZnO bulk growth:

1. Aqueous fluxes at ambient pressure are often based on highly concentrated solutions of KOH and/or other alkaline hydroxides where different zincates such as described by eq. (5) are formed. The handling of the very aggressive hydroxides is not easy, and besides KOH tends to absorb water and carbon dioxide from air. Resulting from such impurities, technical

KOH was found to melt at 140°C, much below the melting point of the pure substance of 243°C. Often silver crucibles are used because this metal is more stable than even platinum. During growth runs of 2 days at 450-480°C transparent ZnO needles with 6 mm length and 0.2 mm diameter could be grown [18]. By adding LiOH to the solvent also single crystalline plates (diameter 5 mm, thickness 0.1 mm) can be obtained [19]. Usually for all such experiments $\vec{c}$ is either the needle axis or the plate normal.

2.  Typical melt fluxes composed e.g. of vanadates and phosphates [20], $MoO_3$, $B_2O_3$ [21], $PbF_2$ or PbO [22] were used for growing ZnO also. Unfortunately, flux grown crystals are always contaminated by solvent, sometimes up to 1% [21], which makes their application e.g. for electronic devices difficult. An interesting variant allows using a self-flux of molten zinc to grow ZnO crystals [23]. Only small crystals could be obtained, but the purity was reported to be higher compared with commercial hydrothermal material.

3.  Hydrothermal solutions are chemically comparable with the alkaline aqueous fluxes that were mentioned in point 1, but have lower hydroxide concentration. Exact data on process parameters are usually undisclosed by the producers. Hydrothermal growth means conditions beyond the critical point of the solvent. For pure water this is $T_c = 373.9$°C, $p_c = 220.6$ bar. In such reactors a polycrystalline feedstock (powder or compacted grains) is heated together with the solvent inside an autoclave in such a way, that a small temperature gradient of ca. $10-15$ K transports the nutrient to the growth zone where seeds are initiating growth. The process is similar to the technology for mass production of α-quartz crystals that was developed at Bell Laboratories after World War II [24] and is nowadays the almost exclusive source of commercial ZnO bulk crystals. Mainly groups in Japan and Russia grow crystals with several inch diameter in platinum or Ti-alloy lined autoclaves with inner diameter of $\geq 200$ mm and length up to several meters [25,26]. The crystallographic quality of hydrothermally grown crystals is often impressive (etch pit density a few 100 cm$^{-2}$ or even lower, FWHM of (0002) as low as 8 arcsec were reported [25]). One should not forget, however, that hydrothermal growth is still a solution growth technology, with the typical incorporation of solvent traces (especially alkaline metals and hydrogen) in the grown crystals.

## RELATED COMPOUNDS: OXIDES OF GALLIUM, INDIUM, TIN

Besides ZnO, oxides of some other metals (typically from the 3$^{rd}$ or 4$^{th}$ main group) are promising materials e.g. for **T**ransparent **C**onductive **O**xide applications [27]. Often these materials are used as polycrystalline layers, but at least for fundamental research single crystals are desirable. Fig. 7 compares stability fields of three relevant materials with that of ZnO. The topmost black dashed line shows $p_{O_2}(T)$ that is delivered by carbon dioxide at ambient pressure. For all metal-oxygen systems only the phase boundary separating the desired oxide from the corresponding metal that is formed for too low $p_{O_2}$ is shown. For the melt growth of the oxides the system must be kept stable near the melting point that is indicated by a short line towards the oxide side of the phase boundary.

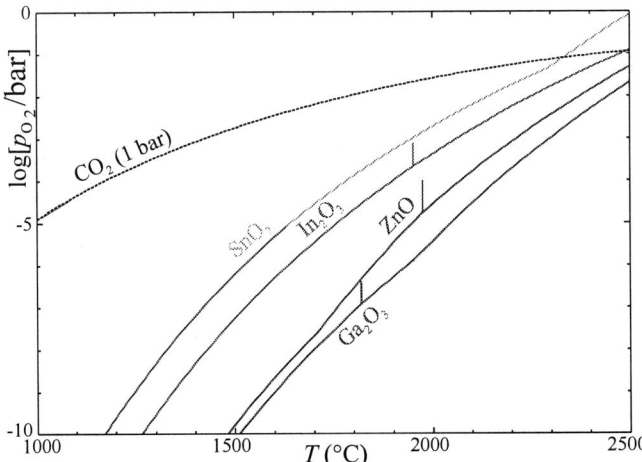

**Fig. 7: Stability fields for the TCO's SnO$_2$, In$_2$O$_3$, ZnO, and Ga$_2$O$_3$ (from top to bottom) compared with the oxygen fugacity that is delivered by pure CO$_2$. The short vertical dashes denote the melting points of In$_2$O$_3$ [28], ZnO [15], and Ga$_2$O$_3$ [29]**

It is obvious that the requirements for the growth of Ga$_2$O$_3$ crystals are somewhat softer, compared with ZnO: the melting point is lower and less oxygen is required to minimize or avoid decomposition. The feasibility of the Czochralski growth for β-Ga$_2$O$_3$ was demonstrated already a decade ago [30] and boules being suitable for the production of 10×10 mm$^2$ wafers were grown recently [29]. The requirements for In$_2$O$_3$ are much harder, because about one order of magnitude more oxygen is required to stabilize this oxide at the melting point, compared with ZnO. Nevertheless, the first truly bulk In$_2$O$_3$ single crystals grown from the melt were demonstrated [28]. Attempts to grow SnO$_2$ crystals from the melt failed so far.

Unfortunately for SnO$_2$ the current knowledge even about the melting temperature is insufficient. Often values around 1630°C are reported [31], but this was proven to be wrong by the current authors. Pure SnO$_2$ powder heated in flowing oxygen in a NETZSCH STA409 to 1650°C showed considerable evaporation but no melting. One can assume that partial oxygen loss of SnO$_2$, and subsequent melting of Sn, resulted in wrong (and too low) melting point data found in literature. Because the stability limit of SnO$_2$ (green curve in Fig. 7) approaches the oxygen partial pressure that is supplied even by pure CO$_2$ it remains questionable whether melt growth of pure tin dioxide from metallic crucibles will ever be possible.

"ITO" (indium tin oxide) is certainly one of the most important TCO's that is used in polycrystalline form e.g. for transparent electrodes in flat panel displays or solar cells. At room temperature this is a mixture of In$_2$O$_3$ saturated with Sn, and SnO$_2$ saturated with In. The solubility limit for both dopants is ≈10% [32]. Only for $T > 1200$°C an intermediate compound with homogeneity range around In$_2$SnO$_5$ is stable. For lower $T$ this phase undergoes decomposition.

It would be desirable to identify systems where such intermediate compounds, or at least solid solutions, are stable or metastable down to room temperature. Already in the case of ideal miscibility the free enthalpy of the solid (also of the liquid) phase decreases then by

$$\Delta G_{\text{mix}} = -RT(x_1 \ln x_1 + x_2 \ln x_2)$$
$$\approx -0.7RT \text{ (at } x_1 = x_2 = 0.5)$$

(7)

The oxygen partial pressure in the system corresponds to a Gibbs free energy

$$\Delta G = -RT \ln p_{O_2} \qquad (8)$$

This mean that under the assumption of ideal mixing the oxygen partial pressure that is necessary to stabilize the condensed phases is lowered by $\ln[p_{O_2}] \approx 0.7$, and hence $\log[p_{O_2}] \approx 0.3$. With other words, the stability limit of an ideal mixture of two oxides such as shown in Fig. 7 would be shifted by ca. 1/3 tick mark downwards. Thus even in cases where pure oxides are not accessible, mixture phases could perhaps be grown.

## CONCLUSIONS

Bulk zinc oxide single crystals can be grown by many crystal growth techniques, but nowadays the hydrothermal technique with growth from alkaline aqueous solutions is mostly utilized. The crystals have up to several inch diameter, combined with good crystallographic quality. It should not be forgotten, however, that this is a solution growth method and that traces of the solvent (alkalines, hydrogen) are always incorporated into the crystal. Especially for hydrogen the severe influence on the electronic properties is widely discussed, and hydrogen seems to be one of the reasons why $p$-type doping of ZnO is so difficult.

It should be kept in mind that nearly all other semiconductor crystals with technical relevance (e.g. Si, GaAs) are grown from melts, and so melt grown ZnO could be an alternative for the standard hydrothermal material. Besides ZnO, the oxides of gallium, indium and tin are prospective TCO materials. At least $Ga_2O_3$ and $In_2O_3$ can now be grown from the melt, whereas $SnO_2$ remains a challenge.

## REFERENCES

1. D. Seyferth, *Organometallics* **20**, 2940 (2001). doi: 10.1021/om010439f

2. D. Klimm, D. Schulz, S. Ganschow, in: P. Bhattacharya, R. Fornari, and H. Kamimura (Eds.), *Comprehensive Semiconductor Science and Technology*, Vol. 3, p. 302, Amsterdam (2011).

3. J. C. Rojo, S. Liang, H. Chen, M. Dudley, *Proc. SPIE* **6122**, 61220Q, doi:10.1117/12.656322.

4. E. Scharowsky, *Z. Phys.* **135**, 318 (1953), *in German*.

5. R. Helbig, *J. Crystal Growth* **15**, 25 (1972), *in German*.

6. D. C. Look, D. C. Reynolds, J. R. Sizelove, R. L. Jones, C. W. Litton, G. Cantwell, W. C. Harsch, *Solid State Communications* **105**, 399 (1998).

7. M. Mikami, T. Eto, J. Wang, Y. Masa, M. Isshiki, *J. Crystal Growth* **276**, 389 (2005).

8. J.-L. Santailler, C. Audoin, G. Chichignoud, R. Obrecht, B. Kaouache, P. Marotel, D. Pelenc, S. Brochen, J. Merlin, I. Bisotto, C. Granier, G. Feuillet, F. Levy, *J. Crystal Growth* **312**, 3417 (2010).

9. J.-M. Ntep, S. Said Hassani, A. Lusson, A. Tromson-Carli, D. Ballutaud, G. Didier, R. Triboulet, *J. Crystal Growth* **207**, 30 (1999).

10. G. Brauer, W. Anwand, D. Grambole, J. Grenzer, W. Skorupa, J. Čížek, J. Kuriplach, I. Procházka, C. C. Ling, C. K. So, D. Schulz, D. Klimm, *Phys. Rev. B* **79**, 115212 (2009).

11. O. Fritsch, *Annalen der Physik* **22**, 375 (1935), *in German*.

12. J. Burmeister, *phys. stat. sol. (b)* **10**, K1 (1965), *in German*.

13. J. E. Nause, *III-Vs Review* **12**, 28 (1999).

14. J. Nause, B. Nemeth, *Semiconductor Science and Technology* **20**, S45 (2005).

15. D. Schulz, S. Ganschow, D. Klimm, *J. Crystal Growth* **296**, 27 (2006).

16. D. Klimm, S. Ganschow, D. Schulz, R. Bertram, R. Uecker, P. Reiche, R. Fornari, *J. Crystal Growth* **311**, 534 (2009).

17. D. Klimm, S. Ganschow, D. Schulz, R. Fornari, *J. Crystal Growth* **310**, 3009 (2008).

18. S. C. Kashyap, *J. Appl. Phys.* **44**, 4381 (1973).

19. H. Hashimoto, F. Hayashi, T. Uematsu, Y. Moriyoshi, *J. Mat. Sci. Lett.* **1**, 4 (1982).

20. B. M. Wanklyn, *J. Crystal Growth* **7**, 107 (1970).

21. K. Oka, H. Shibata, S. Kashiwaya, *J. Crystal Growth* **237-239**, 509 (2002).

22. K. Fischer, E. Sinn, *Crystal Research and Technology* **16**, 689 (1981).

23. S.-H. Hong, T. Sato, M. Mikami, M. Uchikoshi, K. Mimura, Y. Masa, M. Isshiki, *J. Crystal Growth* **311**, 3451 (2009).

24. A. C. Walker, *J. Amer. Ceram. Soc.* **36**, 250 (1953).

25. E. Ohshima, H. Ogino, I. Niikura, K. Maeda, M. Sato, M. Ito, T. Fukuda, *J. Crystal Growth* **260**, 166 (2004).

26. L. N. Dem'yanets, V. I. Lyutin, *J. Crystal Growth* **310**, 993 (2008).

27. E. Fortunato, D. Ginley, H. Hosono, D. C. Paine, *MRS Bull.* **32**, 242 (2007).

28. Z. Galazka, R. Uecker, K. Irmscher, D. Schulz, D. Klimm, M. Albrecht, M. Pietsch, S. Ganschow, A. Kwasniewski, R. Fornari, *J. Crystal Growth*, in print (2011), doi: 10.1016/j.jcrysgro.2011.10.029.

29. Z. Galazka, R. Uecker, K. Irmscher, M. Albrecht, D. Klimm, M. Pietsch, M. Brützam, R. Bertram, S. Ganschow, R. Fornari, *Crystal Research and Technology* **45**, 1229 (2010).

30. Y. Tomm, P. Reiche, D. Klimm, T. Fukuda, *J. Crystal Growth* **220**, 510 (2000).

31. FactSage 6.2 Software and Databases, GTT Technologies Herzogenrath (Germany), www.factsage.com (accessed Nov.4, 2011).

32. H. Enoki, J. Echigoya, H. Suto, *J. Mat. Sci.* **26**, 4110 (1991).

**Mater. Res. Soc. Symp. Proc. Vol. 1394 © 2012 Materials Research Society**
**DOI: 10.1557/opl.2012.258**

## Microstructural, Optical and Electrical Properties Of Post-Annealed ZnO:Al Thin Films

Coralie Charpentier[1,2], Patricia Prod'Homme[1], Loïc Francke[1] and Pere Roca i Cabarrocas[2]

[1]TOTAL S.A., Gas & Power R&D Division Tour Lafayette, 2 place des Vosges La Défense 6, 92400 Courbevoie, France.
[2]LPICM-CNRS, Laboratoire de Physique des Interfaces et Couches Minces, Ecole Polytechnique, 91128 Palaiseau, France.

### ABSTRACT

Aluminum-doped zinc oxide (ZnO:Al) thin films were prepared on glass substrates by radio frequency (RF) magnetron sputtering from a ceramic mixed target ZnO:Al$_2$O$_3$ (1 wt.%) with a power of 250 W. Two series of samples were deposited at room temperature, the first one in pure Ar atmosphere, the second one in Ar/O$_2$ gas mixture. Effects of post-deposition annealing treatments carried out from 400 °C to 500 °C under vacuum and in N$_2$/H$_2$ (5%) atmosphere have been investigated. The influence of these parameters was studied by a detailed microstructural analysis using X-Ray diffraction and Raman spectroscopy. For N$_2$/H$_2$ annealing process, the increase of charge carrier concentration limits the increase of the mobility while after vacuum annealing, an improvement of both electrical and optical properties was observed. The increase of the crystallinity and grain size for ZnO:Al films deposited in Ar/O$_2$ gas mixture could explain their improvements. Resistivity was reduced down to $3.5 \times 10^{-4}$ Ω.cm, for a mobility of 49 cm²/V.s with a vacuum annealing at 450 °C for ZnO:Al deposited in Ar/O$_2$ gas mixture.

### INTRODUCTION

In hydrogenated microcrystalline silicon (μc-Si:H) solar cells, doped and intrinsic layers are packed between a transparent conductive oxide (TCO) front contact and a highly reflective back contact. In PIN configuration, light first enters through the front glass, then goes through the TCO layer and is trapped within the intrinsic silicon layer thanks to the light scattering properties of the TCO layer and reflection on the back contact layer. TCOs are crucial for the improvement of μc-Si:H solar cells. TCO front contact must meet a number of requirements [1-4]. This layer must be highly transparent in the active wavelength range of the absorber layer, between 400 nm and 1100 nm for μc-Si:H, and also highly conductive to avoid ohmic losses. RF-sputtered ZnO:Al is a promising material. The light scattering properties are obtained after deposition by a wet etching step. One way to improve both optical and electrical properties of ZnO:Al film is a post-annealing treatment. During the sputtering process, the influence of the substrate temperature is known to improve the opto-electronic properties of the films [5-6]. Furthermore, hydrogen acts as shallow donor in zinc oxide [7] and can improve the conductivity of ZnO films by acting as an effective doping element [8-12]. The use of a ceramic ZnO:Al$_2$O$_3$ target enables a deposition process in pure argon atmosphere. Adding a small amount of oxygen increases the transmittance of ZnO:Al in the visible range but degrades its electrical properties [13]. Many studies have discussed annealing in vacuum [14-16] and hydrogen incorporation [8-12,17] but a deeper understanding of the relationship between microstructural and electrical properties is still required. In this work, the influence of the annealing temperature and the annealing environment on the microstructural, optical and electrical properties of ZnO:Al thin films prepared by RF magnetron sputtering on glass substrates have been studied. A wide variety of characterization techniques is used, including NIR/VIS/UV spectrophotometry and Hall effect measurements. A detailed analysis of X-Ray diffraction complemented by Raman spectroscopy was performed to obtain additional information on microstructure, crystallinity and defects in the films. These films are quite smooth before the wet etching step and could be used also as contact layers in other thin-film photovoltaic platforms such as CIGS or CdTe.

## EXPERIMENTS

ZnO:Al thin films were deposited on Alkaline Earth Boro-Aluminosilicate glass substrates (thickness of 1 mm) at room temperature by RF magnetron sputtering using a $ZnO:Al_2O_3$ (1 wt.%) ceramic target of 152 mm in diameter, and a RF power of 250 W. The base pressure was $4 \times 10^{-5}$ Pa and the working pressure was controlled by the flow rates of Ar and gas mixture $Ar+3\%O_2$, regulated at 0.12 Pa for the two sets of samples. Depositions were carried out in pure argon atmosphere for the first set of samples and in a mixture of Ar and $Ar+3\%O_2$ for the second set of samples in order to have 2% of $(Ar+3\%O_2)$ in the reactive mixture. All studies were carried out on $1000 \pm 100$ nm thick films.

The as-deposited ZnO:Al thin films were subsequently heat-treated for 90 minutes under vacuum and $N_2/5\%H_2$ atmosphere by varying the annealing temperature from 400 °C to 500 °C. Under vacuum, the heat treatment was carried out at a working pressure of $4 \times 10^{-2}$ Pa from a base pressure of $5 \times 10^{-3}$ Pa. The heating ramp was 15 °C/min and the samples were held at the set temperature for up to 90 minutes before cooling down at 3 °C/min. In the $N_2/5\%H_2$ atmosphere, the set temperature was obtained in 5 minutes, and upheld during 90 minutes before a decrease of the temperature in 15 minutes. The X-Ray diffraction (XRD) diagrams were recorded with a PANalytical X'Pert powder diffractometer equipped with a Cu Kα radiation ($\lambda = 1.5418$ Å). The Raman experimental setup consists of a high-resolution (0.1 cm$^{-1}$) Raman spectrometer (Labram HR800 from HORIBA Jobin Yvon) in a confocal microscope backscattering configuration. The accommodate objective used in this study was a 100× (NA = 0.9) objective from Olympus. A tunable 514 nm wavelength Ar laser was employed in this confocal configuration. For all the Raman Stokes analyses the collection time was 10 s with six consecutive spectrum accumulations. A HMS5000 from Microworld was used to evaluate the Hall mobility, carrier concentration and resistivity of the films by Hall effect, and total transmittance measurements were carried out using a Perkin Elmer Lambda 950 UV/VIS/NIR spectrophotometer.

## RESULTS

### Electrical properties

The electrical properties of the ZnO:Al thin films as a function of the annealing temperature under vacuum atmosphere or in 5% hydrogen atmosphere is shown in Fig. 1. Under vacuum environment, an improvement of the electrical properties is observed at an annealing temperature of 400 °C for the ZnO:Al film deposited in pure Ar atmosphere, and 450 °C for the sample deposited in $Ar/O_2$ gas mixture. In 5% hydrogen atmosphere, the highest mobility is observed at 400 °C. For both cases, a degradation of the electrical properties is observed at 500 °C. The charge carrier concentration increases in hydrogen-rich atmosphere annealing, while it remains stable under vacuum annealing. Comparatively the electrical properties are the best under vacuum annealing for ZnO:Al deposited in $Ar/O_2$ gas mixture, with a maximum mobility of 49 cm²/V.s, and a minimum resistivity of $3.5 \times 10^{-4}$ Ω.cm at 450 °C.

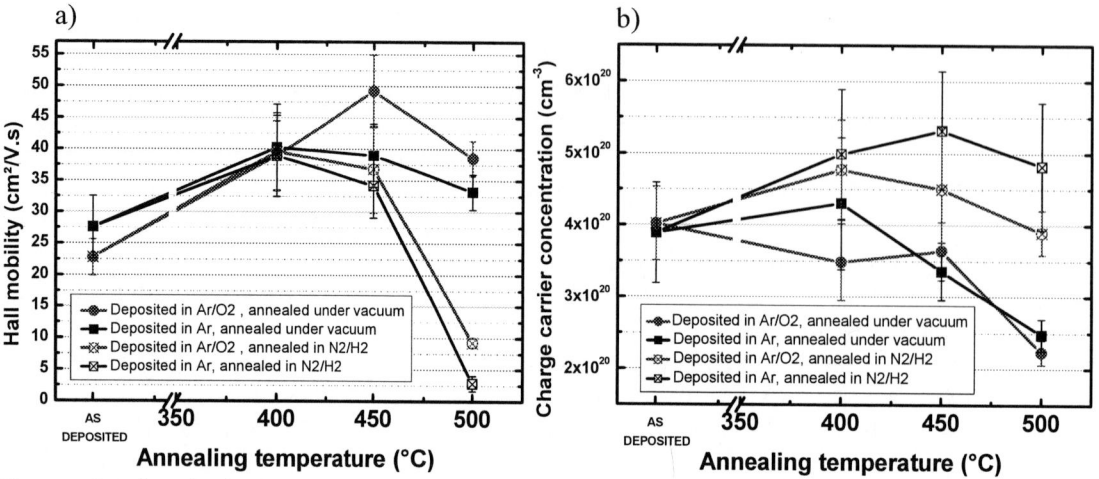

**Figure 1.** Electrical properties as function of the annealing temperature under vacuum (filled symbols) and in $N_2/H_2$ atmosphere (open symbols) for ZnO:Al deposited in pure Ar atmosphere (squares) and in Ar/$O_2$ gas mixture (circles). (a) Hall mobility. (b) Carrier concentration.

## Optical properties

Fig.2 shows the optical transmittance in the 300 nm - 1500 nm wavelength range for the as-deposited and annealed ZnO:Al thin films deposited in Ar/$O_2$ gas mixture, the films deposited in pure Ar atmosphere exhibiting the same optical behavior. All films exhibit an increase from 75% up to 83% of the transmittance in the 400 nm - 1000 nm range after the annealing step. A blue shift of the optical absorption edge is observed with increasing the annealing temperature, slightly for vacuum annealing but strongly for $N_2/H_2$ atmosphere annealing. This optical band gap widening is due to the shift of the Fermi level in the conduction band by the Burstein-Moss effect, and is consistent with the increase of the charge carrier concentration for ZnO:Al films annealed in $N_2/H_2$ atmosphere. [8-10,14,16-17] A reduced fringe profile is observed after the $N_2/H_2$ annealing at 500 °C, due to an increase of the roughness of the film.

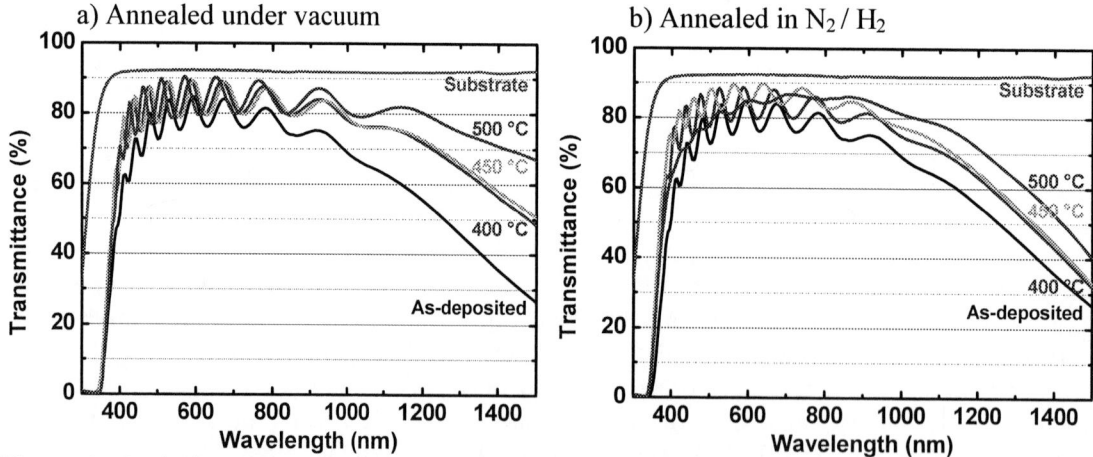

**Figure 2.** Optical transmittance spectra of ZnO:Al thin films deposited in Ar/$O_2$ gas mixture, as deposited and annealed under vacuum (a), and in $N_2/H_2$ atmosphere (b).

## Microstructural properties

Raman spectroscopy can provide qualitative information about the crystallinity of the films and can be correlated with X-Ray diffraction [18]. The Raman spectra of the annealed ZnO:Al thin films presented in Fig.3 are dominated by the $A_1$-LO vibrations around 578 cm$^{-1}$. This asymmetric band arises from two contributions, a low wavenumber contribution and a high wavenumber contribution. Indeed, the relative area of the disordered (low wavenumber contribution) and crystalline (high wavenumber contribution) peaks enables to compare the degree of crystallinity corresponding to different spectra, revealed by an apparent shift of the LO mode. With increasing annealing temperature, the $A_1$-LO peak is apparently shifted towards higher wavenumbers (until 400 °C or 450 °C depending on the annealing environment). This observation comes from the increase of the high wavenumber contribution compared to the low wavenumber contribution, suggesting an improvement of the crystallinity and a reduction in defects of the ZnO:Al deposited in Ar/O$_2$ gas mixture films between 400 °C and 450 °C. This increase of the crystallinity is correlated with the improvement of the electrical and the optical properties of the films.

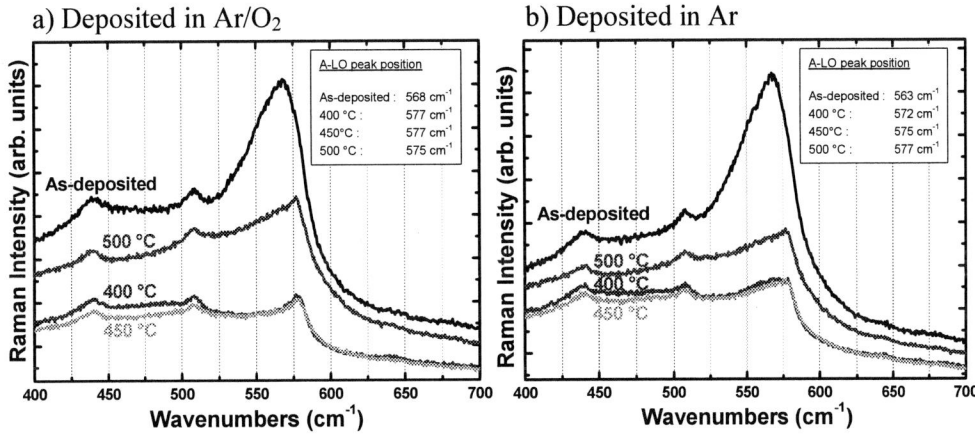

**Figure 3.** Raman spectra of the ZnO:Al thin films deposited in Ar/O$_2$ gas mixture (a) and in Ar atmosphere (b), as deposited and annealed at 400 °C, 450 °C and 500 °C in N$_2$/H$_2$ atmosphere.

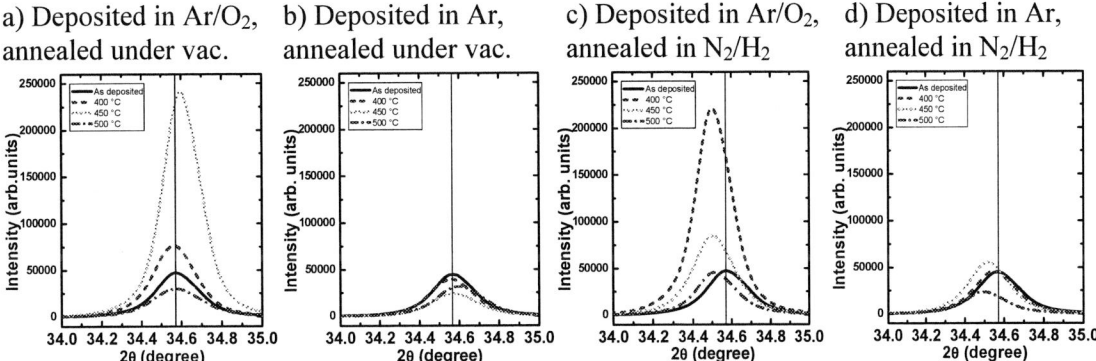

**Figure 4.** XRD diagrams of the ZnO:Al deposited in pure Ar atmosphere (a/c) or in Ar/O$_2$ gas mixture (b/d), as deposited and annealed at 400 °C, 450 °C and 500 °C under vacuum (a/b), and in N$_2$/H$_2$ atmosphere (c/d).

All films exhibit a strong c-axis preferred orientation revealed by the high intensity of the (002) reflection of ZnO:Al Würtzite structure. The XRD diagrams (Fig.4) show a strong increase of the crystallinity for ZnO:Al thin films deposited in Ar/$O_2$ gas mixture, annealed at 450 °C under vacuum, and samples annealed at 400 °C in $N_2$/$H_2$ atmosphere. This improvement of the crystallinity is revealed by the increase of the intensity of the (002) peak, and the decrease of the full width at half minimum (FWHM), corresponding to a higher grain size or a lower mosaicity. No clear increase of crystallinity of the ZnO:Al samples deposited in pure argon atmosphere is observed in XRD diagrams after annealing process. The (002) peaks are shifted towards lower 2θ for ZnO:Al samples annealed in $N_2$/$H_2$, revealing a lattice relaxation.

## DISCUSSION

In both cases, under vacuum or in $N_2$/$H_2$ atmosphere, the annealing process improves the electrical and optical properties of ZnO:Al films, between 400 °C and 450 °C. However, the microstructural, optical and electrical studies do not show the same behavior for each case. The highest improvement is observed for the ZnO:Al thin film deposited in Ar/$O_2$ gas mixture, and post-annealed in vacuum at 450 °C.

First, the deposition atmosphere has an important influence on the microstructure of the film after annealing as shown in Raman spectra (Fig.3) and in XRD diagrams (Fig.4). A high increase of the crystallinity with annealing is observed for ZnO:Al thin films deposited in Ar/$O_2$ gas mixture, while the microstructure does not change for samples deposited in pure argon atmosphere. The increase in optical transmittance is consistent with this improvement of the crystallinity of the films and the increase of the grain size (Fig.2). Thereby, after annealing, the improvement of the electrical properties is higher for ZnO:Al deposited in Ar/$O_2$ gas mixture than for samples deposited in pure argon atmosphere as seen in Fig.1, improvement attributed to the decrease in scattering at grain boundaries and improved crystallinity. [8,10,17]

Secondly, the annealing atmosphere does not seem to have an effect on the crystallinity of the film. For $N_2$/$H_2$ annealing atmosphere, the incorporation of hydrogen inside the lattice leads to a distortion of the ZnO Würtzite structure. It results in a lattice relaxation and an increase of $d_{(002)}$ interplaner spacing, revealed by the shift towards lower angles of the (002) diffraction peaks for ZnO:Al annealing in hydrogen-rich environment as seen in Fig.4. The annealing atmosphere influences the charge carrier concentration. A decrease of the resistivity is due to an increase of the mobility and/or carrier concentration. The electrical properties are better after vacuum annealing than after a hydrogen-rich annealing. During vacuum annealing, the charge carrier concentration remains stable, while it increases with $N_2$/$H_2$ annealing because of the incorporation of hydrogen into the ZnO:Al layer, acting as dopants [11,12]. The relationship between carrier concentration and mobility has been discussed by Bellingham et al. [19], Minami [20] and Ellmer [21]. They found that the ionized impurity scattering limits the maximum mobility through a linear dependence μ-N, for carrier concentration in the range of $10^{19}$ $cm^{-3}$ - $10^{21}$ $cm^{-3}$ for Bellingham, in the range of $10^{20}$ $cm^{-3}$ - $10^{21}$ $cm^{-3}$ for Minami, and for N > $10^{20}$ $cm^{-3}$ for Ellmer. For $N_2$/$H_2$ annealing, the high carrier concentration limits the carrier mobility, which is consistent with the mobility limit of 40 cm²/V.s for N > $5 \times 10^{20}$ $cm^{-3}$ given by Ellmer [21]. Further measurements as temperature-dependent Hall Effect measurement have to be done to confirm this hypothesis. Thereby, the best properties observed for the ZnO:Al thin film deposited in Ar/$O_2$ gas mixture, and annealed in vacuum at 450 °C could be explained by an improvement of the crystallinity and increase in grain size, and by the moderate carrier concentration.

## CONCLUSIONS

This study is focused on the influence of a post-annealing step on electrical, optical and microstructural properties of ZnO:Al thin films deposited at room temperature. The effects of the deposition atmosphere, annealing atmosphere and temperature were studied. The electrical optimum is achieved under vacuum annealing at 450 °C, with a maximum mobility of 49 cm$^2$/V.s, and a minimum resistivity of $3.5 \times 10^{-4}$ $\Omega$.cm. Depositions realized in Ar/O$_2$ gas mixture show the best improvement of the crystallinity after annealing. The N$_2$/H$_2$ annealing leads to an increase of the carrier concentration, which limits the mobility by ionized impurity scattering. These thin films are used as front contact layers for μc-Si:H solar cells and could be appropriate for other thin film photovoltaic platforms such as CIGS or CdTe.

## REFERENCES

1. B. Rech, T. Repmann, M.N. Van Den Donker, M. Berginski, T. Kilper, J. Hüpkes, S. Calnan, H. Stiebig, and S. Wieder, *Thin Solid Films* **511**, 548 (2006).

2. W. Beyer, J.Hüpkes, and H. Stiebig, *Thin Solid Films* **516**, 147 (2007).

3. J. Springer, B. Rech, W. Reetz, J. Muller, and M. Vanecek, *Sol. Energy Mater. Sol. Cells* **85**, 1 (2005).

4. J. Müller, B. Rech, J. Springer, and M. Vanecek, *Sol. Energy* **77**, 917 (2004).

5. R. Wen, L. Wang, X. Wang, G.H. Yue, Y. Chen and D.L. Peng, *J. Alloys Compd.* **508**, 370 (2010).

6. S. Singh, R.S. Srinivasa, and S.S. Major, *Thin Solid Films* **515**, 8718 (2007).

7. C. Van De Walle, *Phys. Rev. Lett.* **85**, 1012 (2000).

8. B.L. Zhu, J. Wang, S.J. Zhu, J. Wu, R. Wu, D. W. Zeng, and C.S. Xie, *Thin Solid Films* **519**, 3809 (2011).

9. B.Y. Oh, M.C. Jeong, D.S. Kim, W. Lee and J.M. Myoung, *J. Cryst. Growth* **281**, 475 (2005).

10. C. Lennon, R.B. Tapia, R. Kodama, Y. Chang, S. Sivananthan and M. Deshpande, *J. Electron. Mat.* **38**, 1568 (2009).

11. S.J. Tark, M.G. Kang, S. Park, J.H. Jang, J.C. Lee, W.M. Kim, J.S. Lee and D. Kim., *Curr. Appl. Phys.* **9**, 1318 (2009).

12. J.N. Duenow, T.A. Gessert, D.M. Wood, D.L. Young, and T.J. Coutts, *J. Non-Cryst. Solids* **354**, 2787 (2008).

13. K. Ellmer, *J. Phys.D.:Appl. Phys.* **33**, 17 (2000).

14. Y. Kim, W. Lee, D.R. Jung, J. Kim, S. Nam, H. Kim, and B. Park, *Appl. Phys. Lett.* **96**, 171902 (2010).

15. F. Ruske, M. Roczen, K. Lee, M. Wimmer, S. Gall, J. Hüpkes, D. Hrunski, and B. Rech, *J. Appl. Phys.* **107**, 013708 (2010).

16. Z. Ben Ayadi, L. El Mir, K. Djessas, S. Alaya, *Thin solid films* **517**, 6307 (2009)

17. W. Yang, Z. Wu, Z. Liu, A. Pang, Y.L. Tu, and Z.C. Feng, *Thin Solid Films* **519**, 31 (2010).

18. C. Charpentier, P. Prod'homme, I. Maurin, M. Chaigneau and P. Roca i Cabarrocas, *EPJ Photovoltaics* **2**, 25002 (2011)

19. J.R. Bellingham, W.A. Phillips, C.J. Adkins, *J. Mater. Sci. Lett.* **11**, 263 (1992).

20. T. Minami, *MRS Bull.* **25**, 38 (2000).

21. K. Ellmer, *J.Phys.D* **34**, 3097 (2001).

**Mater. Res. Soc. Symp. Proc. Vol. 1394 © 2012 Materials Research Society**
**DOI: 10.1557/opl.2012.417**

## ZnS$_x$O$_{1-x}$ Films Grown on Flexible Substrates

Jesse Huso[1], Hui Che[1], John L. Morrison[1], Dinesh Thapa[1], Michelle Huso[1], Stanley Rhodes[1], Brianna Blanchard[1], Wei Jiang Yeh[1], M. D. McCluskey[2] and Leah Bergman[1]

[1] Physics Department, University of Idaho
Moscow, ID 83844-0903, U.S.A.
[2] Department of Physics and Materials Science Program, Washington State University
Pullman, WA 99164-2814, U.S.A.

### ABSTRACT

Bandgap engineered ZnS$_x$O$_{1-x}$ films were grown on Fluorinated Ethylene Propylene (FEP) substrates and analyzed using transmission spectroscopy. FEP is considered as a potential substrate for application in flexible electronics and semiconductor films.

### INTRODUCTION

ZnO has been widely investigated for potential use as a blue and UV light source. Due to its relatively benign chemistry ZnO also presents the possibility of reducing the usage of toxic materials in technology. However, the ability to engineer the bandgap of a ZnO based material is essential to the creation of such devices. The alloy system Mg$_x$Zn$_{1-x}$O has been known for some time to allow the bandgap of ZnO based materials to be moved deeper into the UV range, however there are few options for achieving a decrease of the bandgap into the visible range. While a decrease of the bandgap into the visible range is possible, achieving such a change typically involves the use of toxic materials such as cadmium[1,2]. There are alternatives however.

It has been show that highly lattice mismatched alloy systems such as ZnO$_{1-x}$Se$_x$ [3] and ZnS$_x$O$_{1-x}$ [4] display extreme bowing parameters that can result in bandgaps well below those of either of the end members. In this work we consider the ZnS$_x$O$_{1-x}$ system. The end member ZnS is a compound widely found in nature as the mineral sphalerite, and is quite benign chemically. While both ZnO and ZnS have bandgaps in the UV range, 3.34 eV and 3.84 eV [5] respectively, the large bowing parameter allows the bandgap of ZnS$_x$O$_{1-x}$ to reach well into the visible range. Consequently ZnS$_x$O$_{1-x}$ has been suggested as a possible route for achievement of bandgap reduction of ZnO based materials[5] without the use of highly toxic materials such as cadmium.

In addition, with ever more demanding requirements for sensor and display applications, flexible materials are poised to make a very significant impact in many areas. Among other advantages of flexible materials are the ability to fabricate devices in large volumes using roll-to-roll technology[6], the ability to conform to surfaces, potential to be waterproofed, and the potential to remain operable even after repeated bending[7]: conditions that would destroy most electronic devices. However, growing useful materials on flexible substrates is often challenging due in part to the low melting points and poor chemical resistance of many flexible substrate materials. Furthermore, many plastics are opaque in the UV range, severely limiting their usefulness in the UV range. However, progress is being made in these areas. Previous work demonstrated ZnO and MgZnO films grown on fluorinated ethylene propylene (FEP) substrates and investigated their optical properties in the UV range[8]. FEP is highly transparent in the visible and well into the UV range, and in addition is very resistant to chemical attack making

FEP a potential candidate for flexible optical devices in the visible and UV range. This work demonstrates $ZnS_xO_{1-x}$ films grown on FEP as a step toward application of ZnO based flexible materials in the visible range.

**EXPERIMENT**

The $ZnS_xO_{1-x}$ films were grown by reactive sputtering utilizing a custom built RF magnetron sputtering system using a ceramic ZnS target and argon as the sputtering gas. Controlled amounts of oxygen were introduced during the growth process to create the $ZnS_xO_{1-x}$. The films were grown on commercially available FEP substrates. The film composition was measured by energy-dispersive X-ray spectroscopy (EDS). The transmission spectra were obtained using a Cary 300 UV-Vis transmission system operating in double beam mode.

**DISCUSSION**

Fluoropolymers exhibit many desirable properties such as low coefficients of friction, resistance to chemical attack, heat-sealability and high melting points. FEP is a fluoropolymer closely related to poly(tetrafluoroethylene) (PTFE)[9] and has many properties that are similar to PTFE. In addition FEP displays a high degree of transparency even well into the UV range making FEP a good choice as a substrate or capping layer for UV optical devices[8]. Table I lists some relevant physical properties of FEP with PTFE for comparison while Figure 1 shows the structural differences of FEP and PTFE.

**Table I**. Some typical physical properties of FEP with PTFE for comparison purposes.

| Property (units) | FEP | PTFE |
|---|---|---|
| Folding endurance (cycles) [10] | 5-80 x $10^3$ | > $10^6$ |
| Melting point (°C) [11] | 268 | 326 |
| Coefficient of friction [11] | 0.16 | 0.05 |
| Continuous use temperature (°C) [11] | 204 | 260 |

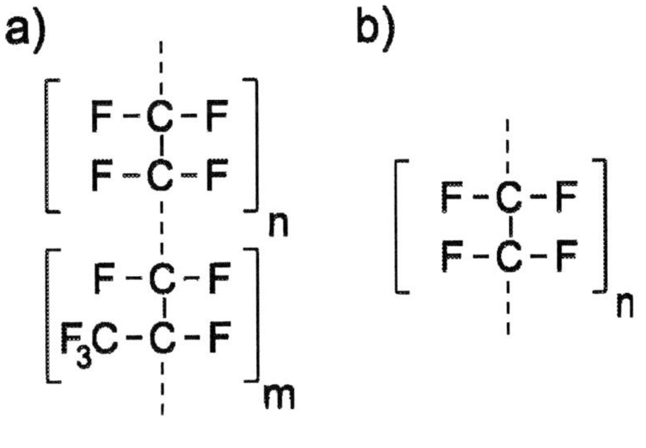

**Figure 1**. The structures of a) FEP and b) PTFE showing their close resemblance, where *n* and *m* are integers. After Crosby *et al.*[11].

## RESULTS

Two $ZnS_xO_{1-x}$ films were grown on FEP substrates and analyzed. EDS measurements of the $ZnS_xO_{1-x}$ films showed that the films had the compositions $ZnS_{0.16}O_{0.84}$ and $ZnS_{0.76}O_{0.24}$. Figure 2a shows a photograph of the produced films and Figure 2b shows a representative SEM image of the films. Based on previous work, the films are estimated to be approximately 3.7 μm thick[8]. The films were found to be polycrystalline and highly flexible with no apparent deterioration in their optical properties after bending.

**Figure 2.** a) Photograph of the $ZnS_{0.16}O_{0.84}$ showing the flexibility and transparency of the films. b) Representative SEM image of the $ZnS_{0.16}O_{0.84}$ indicating the film is polycrystalline.

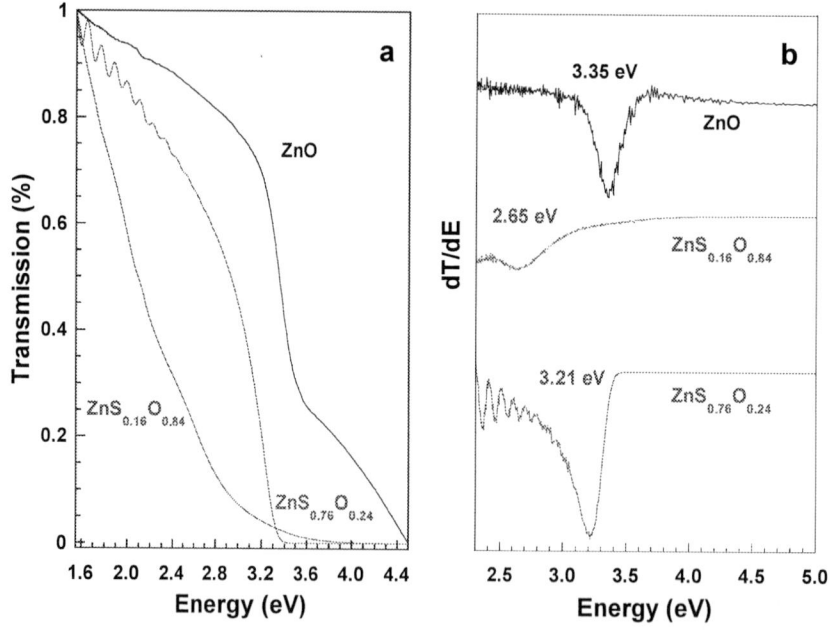

**Figure 3.** a) Transmission plots of $ZnS_{0.76}O_{0.24}$ and $ZnS_{0.16}O_{0.84}$ with data for a ZnO film from Ref. 8 for comparison. b) Plot of the first derivative of the transmission spectra

showing the change of the bandgap energy with addition of sulfur into the crystal lattice.

The optical properties of the $ZnS_{0.16}O_{0.84}$ and $ZnS_{0.76}O_{0.24}$ films were investigated using transmission spectroscopy, with the resulting spectra shown in Figure 3a and for comparison purposes, data from Ref. 8 for a pure ZnO film also grown on FEP is included as well.

To extract the value of the bandgap from the acquired transmission spectra, an inflection point analysis was performed. At the value of the bandgap energy the first derivative of the transmission with respect to energy displays a minimum [12-14]. To find the bandgaps of the films, the first derivative of the transmission was plotted against energy and the results are presented in Figure 3b. The differentiated spectra show the characteristic minima at the bandgap energy and display a composition dependent shift. The bandgap data of the films were found to correlate well with the bandgap values predicted by Moon *et al.*[5]. A summary of the data is shown in Figure 4 where the predicted values of Moon *et al.* have been included for comparison.

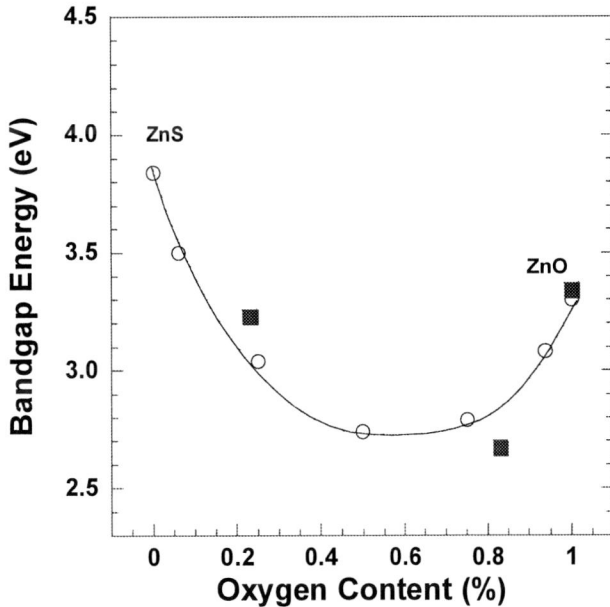

**Figure 4**. Summary of data from Figure 3 with theoretical data for comparison. Data from this work (squares) shows good agreement with theoretical data from Moon *et al.* (circles). The curve is a guide to the eye.

## CONCLUSIONS

Flexible $ZnS_xO_{1-x}$ alloy films have been achieved on FEP substrates through the use of reactive RF sputtering and found by transmission spectroscopy to exhibit reduced bandgaps similar to theoretical predictions. The combination of optical properties, chemical properties and physical properties of FEP and ZnO based materials suggest a route toward possible ZnO based flexible devices.

## ACKNOWLEDGMENTS

This research was supported by the U.S. Department of Energy, Office of Basic Energy

Science, Division of Materials Science and Engineering under Award DE-FG02-07ER46386. The authors also acknowledge the Center of Materials Characterization at the University of Idaho.

**REFERENCES**

1. L. M. Kukreja, S. Barik and P. Misra, *J. Cry. Growth* **268**, 531 (2004).
2. T. Makino, Y. Segawa, M. Kawasaki, A. Ohtomo, R. Shiroki, K. Tamura, T. Yasuda and H. Koinuma, *Appl. Phys. Lett.* **78**, 1237 (2001).
3. M. A. Mayer, D. T. Speaks, K. M. Yu, S. S. Mao, E. E. Haller and W. Walukiewicz, *Appl. Phys. Lett.* **97** 022104 (2010).
4. A. Polity, B. K. Meyer, T. Krämer, C. Wang, U. Haboeck, A. Hoffmann, *Phys. Stat. Sol. (a)* **203**, 2867 (2006).
5. C.-Y. Moon, S.-H. Wei, Y. Z. Zhu and G. D. Chen, *Phys. Rev. B* **74**, 233202 (2006).
6. M. Pagliaro, R. Ciriminna, G. Palmisano, *ChemSusChem* **1**, 880 (2008).
7. R.-H. Kim, D.-H. Kim, J. Xiao, B. H. Kim, S.-I. Park, B. Panilaitis, R. Ghaffari, J. Yao, M. Li, Z. Liu, V. Malyarchuk, D. G. Kim, A.-P. Le, R. G. Nuzzo, D. L. Kaplan, F. G. Omenetto, Y. Huang, Z. Kang and J. A. Rogers, *Nature Materials* **9**, 929 (2010).
8. J. Huso, J. L. Morrison, H. Che, J. P. Sundararajan, W. J. Yeh, D. McIlroy, T. J. Williams and L. Bergman, *Journal of Nanomaterials 2011*, 691582.
9. PTFE is perhaps better known by its trade name Teflon.
10. DuPont, www2.dupont.com/Teflon_Industrial/en_US/tech_info/techinfo_compare.html, Retrieved November 14, 2011.
11. J. M. Crosby, C. A. Carreno and K. L. Talley, *Polymer Composites* **3**, 97 (1982).
12. J. I. Pankove, *Optical Processes in Semiconductors*, 1st ed. (Prentice-Hall, Englewood Cliffs, 1971) pg. 36.
13. R. Viswanatha, S. Chakraborty, S. Basu and D. D. Sarma, *J. Phys. Chem. B* **110**, 22310 (2006).
14. G. Sinha, K. Adhikary and S. Chaudhuri, *J. Crystal Growth* **276**, 204 (2005).

Mater. Res. Soc. Symp. Proc. Vol. 1394 © 2012 Materials Research Society
DOI: 10.1557/opl.2012.434

# Intrinsic Paramagnetic Defects in Zirconium and Hafnium Oxide Films

Robert N. Schwartz,[1,2] Heinrich G. Muller,[1] Paul M. Adams,[1] James D. Barrie,[1] and Ronald C. Lacoe[1]

[1] Electronics and Photonics Laboratory, The Aerospace Corporation, El Segundo, CA 90245, U.S.A.
[2] Department of Electrical Engineering, University of California, Los Angeles, Los Angeles, CA 90024, U.S.A.

## ABSTRACT

Thin films of zirconium oxide ($ZrO_x$) and hafnium oxide ($HfO_x$) were rf sputtered onto fused silica substrates in an oxygen rich argon environment. Pure zirconium and hafnium targets were used, and the oxygen partial pressure was varied to control the oxygen stoichiometry. Measurement of the EPR characteristics of the $ZrO_x$ films indicated two peaks corresponding to two orientations of the magnetic field. This anisotropic response suggested the films were polycrystalline with a preferred orientation. This was confirmed by XRD pole figures. The measured $g$-values for the $ZrO_x$ films were less than the free-spin $g$-value, indicating the defects corresponded to electron traps. It was further shown that the lower the oxygen partial pressure during deposition, the larger the EPR response, strongly suggesting the traps correspond to oxygen vacancies in $ZrO_x$. Hafnium oxide thin films were also characterized by EPR. The EPR measurements indicated the presence of a single resonance peak, suggesting these films were polycrystalline without a preferred orientation or amorphous. XRD measurements confirmed that the $HfO_x$ films were amorphous. The $g$-value for these films was greater than that the free-spin value, indicating the presence of possibly self-trapped oxygen hole centers. These results will be discussed in the context of prior experimental and theoretical work on these systems.

## INTRODUCTION

The scaling of complementary metal-oxide-semiconductor (CMOS) transistors requires the replacement of $SiO_2$ gate oxide by higher dielectric constant materials in order to avoid excessive gate leakage currents [1]. High-κ gate dielectrics, such as $ZrO_2$ and $HfO_2$ and their alloys with $SiO_2$, are currently considered as a practical solution for integrating high-κ materials in future electronic devices [2]-[5]. It is well established that intrinsic point defects, such as oxygen vacancies [3],[4],[6],[7], play an important role in the reliability of high-κ dielectrics and pose one of the fundamental limits for use of high-κ dielectrics in silicon metal – oxide semiconductor field effect transistors [8],[9]. These materials are also used for fabricating anti-reflective dielectric coatings for solar panels used in space-craft applications. The density of defects in these new materials is much greater than found in $SiO_2$. The challenge is to understand the nature of defects in these materials and to evaluate their impact on device performance and reliability.

In recent years, oxides of Group IVA transition metals, Ti, Zr, and Hf have become some of the most investigated materials. It is well established that material properties are strongly influenced by the presence of intrinsic point defects/color centers. Therefore, an in-depth understanding of point defects, at the atomic level, is crucial for evaluating the performance of a material in specific electrical and optical devices, as well as, for developing material fabrication techniques to eliminate them. Here we report on thin films of zirconium and hafnium oxides deposited under similar conditions on fused silica substrates and investigated by electron paramagnetic resonance (EPR) spectroscopy and x-ray diffraction (XRD) techniques. The measurements provide evidence for self-trapped oxygenic hole-centers in hafnium oxide, where as for zirconium oxide, the dominant intrinsic defects result from electron-trapping at possibly oxygen vacancies.

## EXPERIMENTAL METHODS

All films were deposited on fused silica substrates (Corning 7980) purchased from United Lens Company, Inc. Thin films of $ZrO_X$ and $HfO_x$ used for EPR measurements were rf-sputtered from pure metallic zirconium/hafnium targets in a low-pressure oxygen atmosphere. Films were deposited at different zirconium/hafnium deposition rates with different oxygen partial pressures in order to produce films with varying degrees of oxygen stoichiometry. The substrates are not actively heated or cooled during deposition. Plasma bombardment and the heat of condensation do cause appreciable heating of the growing film, but the temperature during deposition does not exceed ~400 K.

All EPR measurements were carried out at 300 K using an $X$-band homodyne Varian E-Line Century Series spectrometer utilizing a $TE_{102}$ rectangular microwave cavity with 25 kHz magnetic-field modulation. The spectrometer is equipped with a Bruker B-H15 magnetic-field controller and the applied magnetic-field values were calibrated by means of a proton magnetic-resonance gaussmeter. To achieve a reasonable signal-to-noise ratio, all spectra reported here were signal averaged over five scans. In addition, for all spectra reported below the following spectrometer settings were used: *microwave power* = 1 mW and *modulation amplitude* = 3.2 G. The estimated uncertainty for the reported $g$-factor values, measured at zero-crossing, is in the range of $\pm$ 0.001 - $\pm$ 0.0004 depending on the line shape of the resonance. The sample disks substrate plus film were cut into rectangular parallelepipeds with thin-film volume ~7 x $10^{-5}$ $cm^3$. A dual-cavity configuration, in conjunction with a Bruker weak pitch standard, was used to determine the concentration of paramagnetic centers in the films. The X-ray diffraction pole figures were acquired with $CuK_\alpha$ radiation using a PANalytical X'Pert Pro diffractometer equipped with an ATC-3 texture cradle (Schulz reflection method).

## EXPERIMENTAL RESULTS AND DISCUSSION

### ZrO$_x$ Films

EPR spectra for a zirconia film are shown below in Fig. 1. The two significant features of this data are: 1) the EPR signal is anisotropic since the resonance varies with orientation of the magnetic field relative to the surface normal; and 2) the $g$-factor of the resonances (*not principal values of the g-tensor*) are less than the free-spin value ($g$ = 2.0023). If the film was truly polycrystalline, one would not expect a difference in EPR signals between the two magnetic field orientations. The fact that the resonances are anisotropic suggests that these films do not have a simple polycrystalline (randomly oriented micro-crystals) structure, but rather polycrystalline with a preferred orientation.

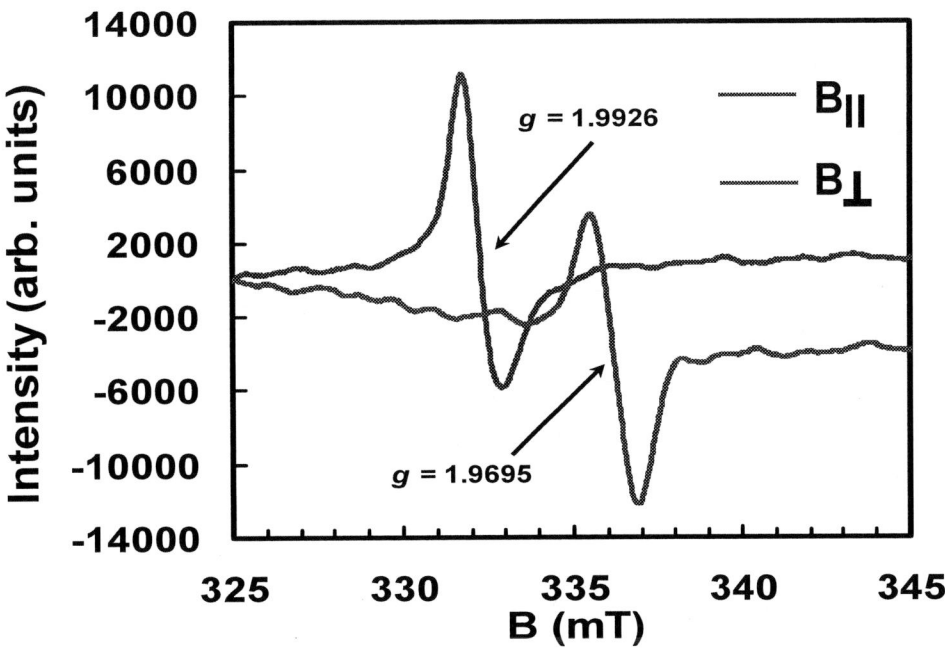

**Figure 1**: Room Temperature EPR spectra of $ZrO_x$ film with the external magnetic field aligned parallel ($B_{\parallel}$, blue) to the surface normal and perpendicular ($B\perp$, green) to the normal (lying in the plane of the film). Spectrometer frequency: $\nu = 9.2673$ GHz. The concentration of paramagnetic centers is in the range of $1 - 3 \times 10^{18}$ centers/cm$^3$ for the as-deposited film.

The observed paramagnetic features appear to be quite stable in the as-deposited films, since the signal intensity as well as line shape in samples fabricated more than a year ago remains unchanged. However, following annealing in both air (308 $^o$C for 3hrs) and forming gas (320 $^o$C for 3 hrs) the EPR spectral features disappeared; subsequent illumination with UV photons (254 nm) did not regenerate the spectra. Below we first address the issue concerning the anisotropic nature of the spectra and then follow with a discussion regarding the nature of the paramagnetic defect center.

X-ray diffraction measurements were made in order to determine the crystalline structure of these films. The diffraction pattern clearly indicated that the films crystallized in the monoclinic phase of zirconia; however, certain peaks that were expected for this phase were missing. These missing lines suggested that the grains had grown with some preferential orientation. To verify this hypothesis, pole figures were taken, which follow any specific lattice spacing with the corresponding wavelength and fixed beam/detector angle, while the sample is tipped and tilted in the beam. This allows one to determine the degree of crystalline alignment about a specific direction.

The peaks investigated were the $\bar{1}11$, *111*, and *003* reflections of the monoclinic zirconia lattice. A resulting pole figure is the standard projection of the peak intensity over the entire forward half sphere. A homogeneous film without preferential orientation would appear as a simple 'disk' with uniform peak intensity over all angles. The pole figures found for the three orientations of a single film are displayed below in Fig. 2. Each pole figure represents the probability of a reflection occurring in any direction of a stereographic projection above the respective film.

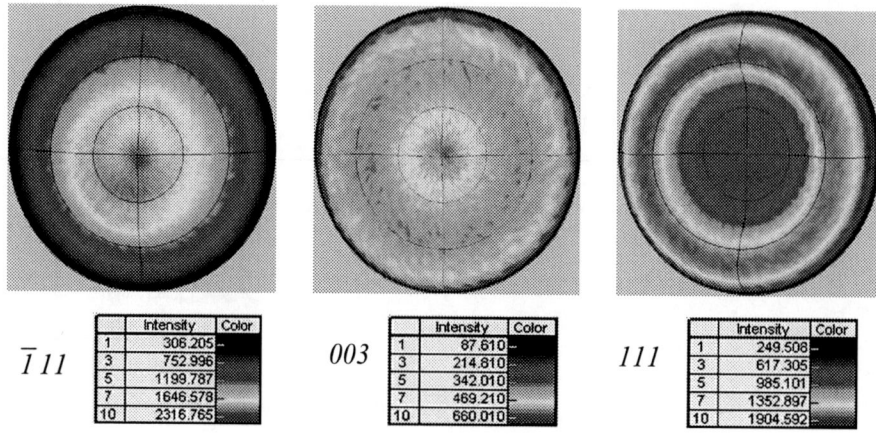

**Figure 2**: Pole figures: left panel: ($\bar{1}11$); $2\theta = 27.75°$, middle panel: (003); $2\theta = 54.66°$, right panel: (111); $2\theta = 31.25°$. The color scale has units of counts.

The resulting pole figures clearly reveal the anisotropic nature of the preferred growth orientation of this film. The $\bar{1}11$ layers of this film grew preferentially parallel to the substrate surface, while the $111$ orientations, which are almost at right angles with the $\bar{1}11$, are forced by this preference to form a circular range around the surface normal. The $003$ pole figure suggests that this orientation has a moderate preference for growing parallel to the substrate, overlaying a large even distribution of grains not following this preference. These results indicate that the inherent anisotropy in the single crystal $ZrO_x$ micro-crystals manifests as an anisotropy in the observed EPR.

From the measured EPR $g$-values one can learn a great deal about the nature of the point defect's atomic structure. The fact that the observed $g$-values in our zirconia films are less than the free-spin $g$-value (2.0023) indicates that the defect giving rise to the resonance corresponds to a trapped electron center; potential trapping sites are oxygen vacancies.

Theoretical calculations of the $g$-value for proposed oxygen-vacancy defects in cubic zirconia have been recently reported by Ramo and coworkers [7]. These authors calculated the electronic structure, optical absorption and magnetic properties ($g$-factors) of oxygen vacancies with charge states $V_O^q$ (V indicates vacancy, subscript O indicates oxygen) spanning the range {q = +2, +1, 0, -1, -2}. Theoretical calculations of $g$-factors of defects, in general, tend to predict values that deviate from those experimentally measured; furthermore, the predicted $g$-factors also vary depending on theoretical approach used [7]. This discrepancy is associated with using idealized models of the point defect in the bulk of the material; potential deviations from this idealized picture involve more realistic models of these centers in association with extended defects such as dislocations, grain boundaries, on surfaces, or in cracks and $g$-strain. However, even with these shortcomings (lack of exact agreement between theoretical calculation and measured $g$-factors), they are still extremely useful for interpreting experimental EPR data.

The theoretical calculations of Ramo *et al.* [7] locate the one-electron defect energy levels for oxygen vacancies in the band gap of $ZrO_2$. The calculated $g$-factors for the S = ½ paramagnetic oxygen vacancy centers are all less than the free-spin value and provide supporting evidence that the observed EPR resonances in our as-deposited $ZrO_2$ films are most likely similar in nature to a trapped electron at an oxygen vacancy. A recent experimental paper that is consistent with electron trapped in oxygen vacancies is the EPR study of defect center production by swift electron and heavy ion irradiation in yttria-stabilized $ZrO_2$, which reported $g$-values similar to ones observed in our zirconium films [10].

Further evidence supporting our view that the EPR resonance feature in as-deposited $ZrO_2$ may be related to an oxygen vacancy defect trap is provided by some preliminary results in which the film fabrication processing conditions were varied. Specifically, samples were prepared to assess the influence of growth rate and oxygen content on the defect characteristics of the films. The growth parameters that could be varied were the sputtering rf power and the oxygen partial pressure. Films grown with the lowest oxygen partial pressure displayed the largest defect population. In contrast, films with the lowest defect concentration were grown under conditions with lower rf power (lower growth rate) and highest oxygen partial pressure.

With regard to the annealing experiments introduced above, we recall that the samples were rendered EPR silent following annealing in either air or forming gas. One can speculate that the thermal annealing process, regardless of environmental atmosphere, may be sufficient to induce local structural changes that destabilize this trapping site. However, a more realistic route for quenching the EPR resonance may involve the thermal excitation of electrons from the paramagnetic $V_O^+$ and $V_O^-$ sites followed by recombination with currently unidentified EPR silent defect centers. These annealing experiments are also important because it allows us to rule out the assignment of the observed EPR in our sputtered films to $Zr^{3+}$. Recent work [11], [12] has shown that this intrinsic defect center, which is identified with $g$-values [12] in the range 1.959 - 1.976 is stable (or in fact, increases slightly in concentration) when samples are heated to elevated temperatures. Clearly this is not what is observed in our films.

Finally, it should be noted that $^{91}Zr$, with natural abundance of 11.2%, has a nuclear spin of I = 5/2 and should lead to well-resolved hyperfine features [13] if highly localized unpaired spin density resides on the $Zr^{3+}$ site. The failure to detect spectral components attributable to hyperfine interactions with $^{91}Zr$ is interpreted as evidence supporting the view that the defect giving rise to the $g < 2.0023$ features in our zirconia film correspond to a S = ½ paramagnetic center with highly localized unpaired spin density residing on the oxygen vacancy.

## HfO₂ Films

HfO$_x$ films were also investigated in this work. An X-ray diffraction pattern is shown in Fig. 3. The HfO$_x$ film, deposited on fused silica, was rf-sputtered from a pure metallic hafnium target in a low-pressure oxygen atmosphere. The X-ray data display broad features that indicate that the film is amorphous; this is in contrast with that observed for the ZrO$_x$ films.

Shown in Fig. 4 is an EPR spectrum measured at 300 K of an HfO$_x$ film (0.2 μm thickness) deposited on fused a silica substrate. The main EPR feature at $g$ = 2.0068 exhibits a line shape that is typical of a randomly oriented system with S = ½. As pointed out above the measured $g$-value provides information about the atomic nature of the point defect. The fact that the observed $g$-values in our HfO$_x$ film is greater than the free-spin $g$-value (2.0023) suggests that the point defect center corresponds to a trapped hole center. Candidate defect structures include a hole trapped on a lattice $O^{2-}$ ion (oxygen hole center) or a hole trapped at a site containing an interstitial oxygen atom ($O_2^-$ center).

## X-ray Diffraction: HfO$_x$ Film

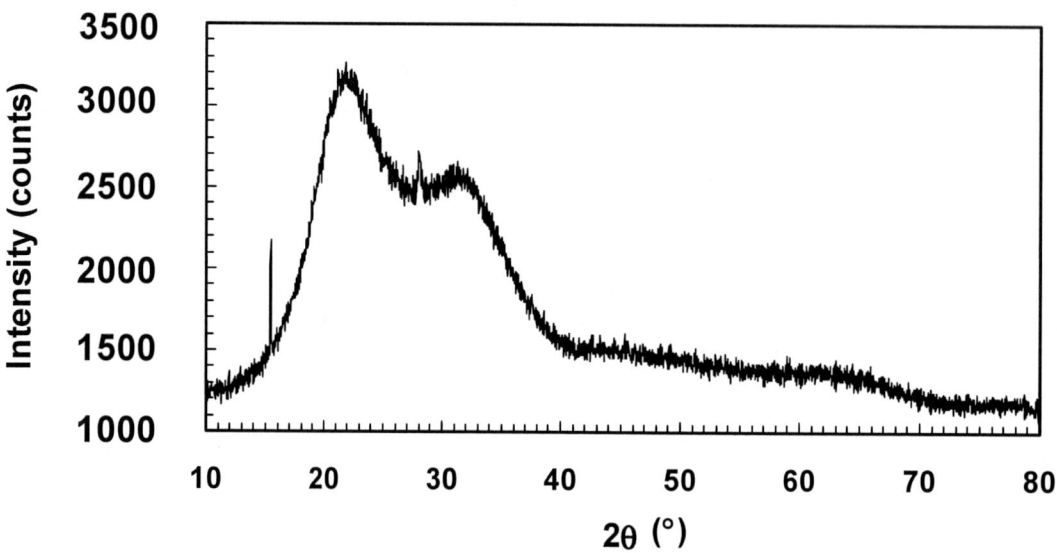

**Figure 3.** X-ray diffraction pattern for a hafnium film. The broad features at ~22° and ~32° are assigned to the fused SiO$_2$ substrate and to amorphous HfO$_x$, respectively. The origin of the sharp feature at ~15° is unknown; the small sharp feature at ~28° is monoclinic polycrystalline HfC

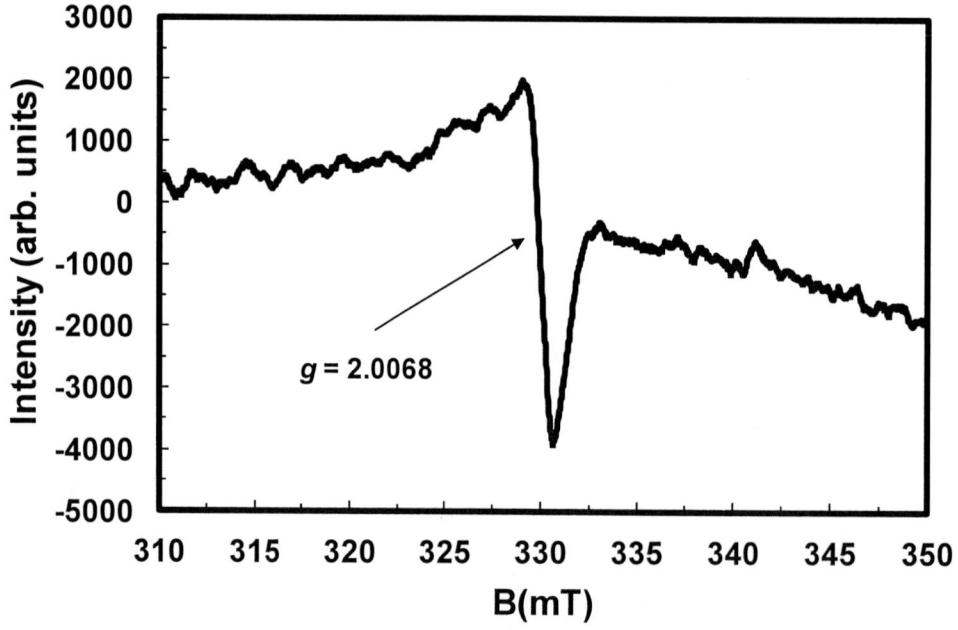

**Figure 4.** Room Temperature EPR spectrum of HfO$_x$ film. Spectrometer frequency: $\nu$ = 9.27067 GHz.

The in-depth electronic structure calculations of Xiong et al [14] provide theoretical evidence for the stability of these types of hole centers in $HfO_2$. Recent experimental evidence for these trapping centers is provided by the EPR studies reported by Lenahan and Conley [15] and Wright and Barklie [12]. It should be emphasized that our results reported here are preliminary results. Further work is in progress to verify that the observed feature in our rf sputtered $HfO_x$ films with $g$-value greater than the free-spin value is an intrinsic defect center.

## SUMMARY

In conclusion, thin films of $ZrO_x$ deposited on fused silica substrates display an anisotropic EPR feature with a $g$-factor less than the free-spin value ($g = 2.0023$). X-ray diffraction data revealed that the films displayed monoclinic morphology and in addition, had a preferential growth normal to the surface of the substrate. As-deposited films annealed in air and forming gas were EPR silent and the paramagnetic center could not be regenerated when illuminated with UV photons. The results of theoretical calculations, along with a series of experiments in which the film fabrication process as well as post-processing conditions could be varied, suggest that the observed paramagnetic centers in rf-sputtered films are most likely associated with trapped electrons at oxygen vacancies in $ZrO_x$. Preliminary EPR results for rf-sputtered $HfO_x$ films indicate that unlike for $ZrO_x$ films that had $g$-values less than the free-spin value, the $HfO_x$ films had g-values greater than the free-spin value, suggesting that the paramagnetic point defects were oxygen hole-traps.

## ACKNOWLEDGMENTS

This work was supported under The Aerospace Corporation's Independent Research and Development Program.

## REFERENCES

1. D.A. Buchanan and S.-H. Lo, Microelectron. Eng. **36**, 13 (1997).
2. G. Wilk, R.M. Wallace, and J.M. Anthony, J. Appl. Phys. **89**, 5243 (2001).
3. J. Robertson, Rep. Prog. Phys. **69**, 327 (2006).
4. K. Xiong, J. Robertson, and S. J. Clark, Phys. Stat. Sol. (B) **243**, 2071 (2006).
5. B.H. Lee, J. Oh, H.H. Tseng, R. Jammy, and L. Huff, Mater. Today **9**, 33 (2006).
6. J.L. Gavartin, D.M. Ramo, A.L. Shluger, G. Bersurker, and B.H. Lee, Appl. Phys. Lett. **89**, 082908 (2006).
7. D. Muñoz Ramo, PV. Sushko, J.L. Gavartin, and A.L. Shluger, Phys. Rev. B **78**, 235432 (2008).
8. A. Kerber, E. Cartier, L. Pantisano, R. Degreaveve, T. Kaueraut, Y. Kim, A. Hou, S. Groesenken, H.E. Maes and U. Schwalke, IEEE Electron Dev. Lett. **24**, 87 (2003).
9. R.J Carter, E. Cartier, A. Kerber, L. Pantisano, T. Shram, S. de Gendt, and M. Heyns, Appl. Phys. Lett. **83**, 533 (2003).
10. J-M. Costantini, F. Beuneu, D. Gourier, C. Trautmann, G. Calas, and M. Toulemonde, *J. Phys.: Condens. Matter* **16**, 3957 (2004).
11. R.N. Schwartz, H. G. Muller, P.D. Fuqua, J.D. Barrie, and R.B. Pan, *Phys. Rev. B* **80**, 134102, (2009).
12. S. Wright and R.C. Barklie, *J. Appl. Phys.* **106**, 103917 (2009).

13. R.F.C. Claridge, K.M. Mackle, G.L.A. Sutton and W.C. Tennant, *J. Phys.:Condens. Matter* **6**, 3429 (1994).
14. K. Xiong, J. Robertson, and S.J. Clark, *Phys. Stat. Sol. B* **243**, 2071 (2006).
15. P.M. Lenahan, and J.F. Conley, Jr., *IEEE Trans. EDM Reliab.* **5**, 90 (2005).

Mater. Res. Soc. Symp. Proc. Vol. 1394 © 2012 Materials Research Society
DOI: 10.1557/opl.2012.435

## Effects of Hydrogen Ion Implantation on Structural Properties of Silver Implantation in ZnO Crystals

Faisal Yaqoob[1] and Mengbing Huang[2,*]
[1] Department of Physics, State University of New York, 1400 Washington Ave, Albany, NY, 12222
[2] College of Nanoscale Science and Engineering, University at Albany, State University of New York, 1400 Washington Ave, Albany, NY, 12222

* Email: mhuang@albany.edu

## ABSTRACT

In this work, we study the effects of implanted hydrogen ions on defect formation and impurity redistribution in ZnO crystals implanted with silver ions. Hydrogen was first implanted at room temperature in ZnO with energy of 30 keV to a dose of $2 \times 10^{16}$ /cm$^2$. The ZnO samples with and without prior H implantation were implanted with Ag ions at four different energies, 30, 75, 150, and 350 keV, to doses $3.3 \times 10^{13}$, $4.2 \times 10^{13}$, $8.3 \times 10^{13}$ and $3.4 \times 10^{14}$ /cm$^2$, respectively, resulting in a uniform concentration profile of Ag from the surface to depth ~ 150 nm. These samples were annealed at temperatures 850-1050°C for 30 minutes in an oxygen gas flow. The distribution of Ag atoms, either aligned or nonaligned along the crystalline directions, were measured by Rutherford backscattering (RBS) combined with ion channeling. Following Ag ion implantation, the damage level in the ZnO lattice, measured along the <10-11> crystalline direction is higher in the sample without H ion implantation than the sample with H. Lattice damage was found to recover faster in the sample without H implantation than the sample with H, e.g., for Zn signals, the normalized RBS yield $\chi_{min}$ for the without H-implanted sample dropped from 27.5% following Ag implantation to 4.3% after annealing at 1050 ˚C, whereas the Zn $\chi$min value for the sample with H implant decreased from 17.6% following Ag implantation to 5.3% after annealing at 1050 ˚C. On the other hand, the $\chi_{min}$ values for the Ag dopants before annealing in the H-implanted sample are the same in the sample without H. Post-Ag-implantation annealing resulted in much higher $\chi_{min}$ values for Ag in the sample with H implant. For the as-implanted samples, 26.6% of the implanted Ag atoms are on substitutional sites in the sample with H, as compared to 30.3% of the implanted Ag being on the substitutional sites in the sample without H. After annealing at 1050 ˚C, the fraction of substitutional Ag is 73.7% in the H-implanted sample, in contrast to the fraction of 61.6% for substitutional Ag in the sample without H implant. Similar to other oxide crystals, H ion implantation and thermal annealing can result in the formation of nanocavities in the ZnO lattice. We discuss these findings in the context of the effects of nanocavities on formation and annihilation of point defects as well as on impurity diffusion and trapping in ZnO crystals.

## INTRODUCTION

ZnO is a promising material for a variety of applications, including ultraviolet light emitters and detectors, high-power and high-frequency electronic devices, sensors, and transparent conductors [1]. Being a wide-band-gap material (3.4eV), it becomes a good candidate for UV optoelectronics [2]. Its transparency to visible light provides opportunities to develop transparent electronics, UV optoelectronics, and integrated sensors all from the same material system. To achieve many of these applications, ZnO still has to overcome the issue of reproducible p-type doping. ZnO is naturally an n-type semiconductor because of the low formation energies native defects like oxygen vacancies and zinc interstitials. Like all other wide band gap semiconductors, (ZnSe, CdS, GaN, etc), it is very difficult to obtain p type doping in ZnO. A number of dopants have been studied, but so far reproducible p type ZnO has been not reported [3-8]. Recently Ag is studied as a potential p type candidate [9-12]. Frist principle calculations have shown very low formation energies for Ag at zinc substitutional sites compared to interstitial sites [13]. Hydrogen in ZnO has been studied widely and is considered to be a donor in ZnO. The experiments of Ip et al. [14, 15] show that the thermal stability of hydrogen is slightly higher when hydrogen is incorporated by direct implantation in ZnO [4]. Also Moutanabbir et al., [16], found vacancy clustering and the formation of nanobubbles when they implanted H in GaN.

In this study, we investigate the effects of H on Ag implantation in ZnO crystals. The idea is to create nano cavities by H implantation and trap the point defects in those cavities and diffuse the Ag atoms to substitutional sites during thermal process. Also the lattice damage and its recovery during annealing are studied.

## EXPERIMENTAL DETAILS

ZnO single crystals a, c and e, each of size of $1 \times 1$ cm$^2$, were used for this experiment. Sample e was kept as the control sample without hydrogen implantation, while the other two samples, a and c, were implanted with hydrogen ions at 30 keV to a dose of $2 \times 10^{16}$ /cm$^2$ from depth region ~ 150 to 300nm. Post H implantation annealing was done on sample c in an argon gas flow at 800 °C for 15 minutes. After this initial step all three samples were ion implanted with silver at four different energies 30, 75, 150, and 350 keV, to doses $3.3 \times 10^{13}$, $4.2 \times 10^{13}$, $8.3 \times 10^{13}$ and $3.4 \times 10^{14}$ /cm$^2$, respectively, resulting in a uniform concentration profile of Ag from the surface to depth ~ 150 nm. During implantation, the samples were tilted by ~ 7° relative to the perpendicular ion beam to minimize channeling effects during implantation. These samples were annealed at temperatures 850-1050 °C for 30 minutes in an oxygen gas flow.

The distribution of Ag atoms, either aligned or nonaligned along the crystalline directions, were measured by Rutherford backscattering (RBS) combined with ion channeling. Helium beam with energy 3.05 MeV, surface charge dose of 8 μC and beam intensity ~ 0.50 nA – 1.00 nA for random and aligned respectively, was used for RBS/Channeling experiment The samples were analyzed along two crystallographic directions, <0001> and <10-11>. When the projectile He beam is steered between the <0001> atomic rows, the helium beam is reflected only from the atoms within that channel. In which case it can miss the Ag atoms in the shadow cone behind a surface atom, and the in-channel fraction of Ag wouldn't reflect the total or most of the Ag concentration within the crystal [20]. Therefore to get a better depth resolution of the impurity atoms, channeling along <10-11> crystallographic direction is also done. In which case, the analyzing He beam can be backscattered from Ag atoms either within the atomic rows on interstitial or on substitutional lattice positions. The samples were also analyzed along random direction. Although pure random direction can only be found for amorphous material, we continuously rotated the sample to get an accurate random spectrum.


**RESULTS**


The Rutherford backscattering channeling spectra, along the <0001> crystallographic direction and random direction, of ZnO following the room temperature ion implantation of Ag in the samples; with H implantation, with H implantation and intermediate annealing in Ar gas and without H implantation are presented in Fig. 1. The RBS/Channeling was done at oxygen resonance energy to observe the displacement of oxygen atoms by ion implantation [20]. The damage created to the crystal lattice by H implantation starting at depth ~ 150nm below the surface as simulated by SRIM depth profile and extends up to ~ 350nm, larger than the SRIM profile which can be correlated to tail in SRIM profile, can be clearly seen in sample a, Fig. 1 (a) and (d). This damage region is annealed out by the intermediate annealing in sample c. As the samples were annealed post Ag implantation, this damage region was annealed out in RBS spectra as shown in Fig. 2 (b). The crystallinity of Zn lattice seems to be holding stronger than the O lattice during implantation, i.e., the $\chi_{min}$ values for zinc signal are 6.31%, 7.34% and 10.50% compared to oxygen $\chi_{min}$ values of 21.4%, 15.4% and 22.9% for samples a, c, and e respectively when measured along the <0001> crystallographic direction. On the other hand the RBS/C yield for zinc signal when measured between the <10-11> atomic rows are 17.6%, 23.1%, and 27.6% compared to oxygen signal values of 58.1%, 57.5% and 53.3% in sample a, c, and e respectively. The crystal structure recovers with annealing as has been observed in literature. The Zn lattice recovers faster and stronger than the O lattice, upon annealing from 850 – 1050 °C. At highest temperature 1050 °C it almost recovers to its virgin sample value. The hydrogen implantation has different effects on zinc and oxygen crystal lattice post silver implantation. The damage caused by the Ag implantation to Zn lattice is smaller in the samples with H, than the samples without H, whereas the oxygen lattice is disturbed more by the Ag silver implantation in the presence of H as shown in RBS spectra in Fig. 1. H implantation has no effect on Ag implantation initially and 78% of the Ag atoms are placed randomly in both cases; with and without H samples, though, the intermediately annealed samples has slightly higher, 79.4%, of randomly placed Ag atoms. As the samples are annealed in oxygen ambient, Ag atoms are diffused out of the implanted layer as can be seen in RBS spectra in Fig. 1 (b) and 2 (b). The Ag diffusion is low in H implanted samples at temperature 850°C and 950°C than the samples without H and intermediately annealed but at 1050 °C Ag diffusion is much faster in the presence of H than the samples without H and intermediately annealed samples.

After Ag implantation, roughly ~ 30.3% of the silver atoms are on substitutional Zn sites ($S_{zn}$) and ~ 69.7% within the <10-11> atomic rows, in the sample without H. *E. Rital et al.*, has found 30% to 42% of the Ag atoms at defect free $S_{zn}$ sites. [17]. They have shown that up to 600°C vacuum annealing the fraction of Ag atoms at $S_{zn}$ increased somewhat to 46%. We have done annealing in oxygen gas flow and got very steady increase in the fraction of Ag at $S_{zn}$ from 30.3% to 61.6%. In samples with H and intermediate annealing there is initially 26.7% of Ag atoms on substitutional Zn sites which becomes maximum, 59.2% at the highest annealing temperature. In samples with H implantation the maximum number ~ 73.7% of Ag atoms on zinc substitutional is achieved after annealing at 1050°.

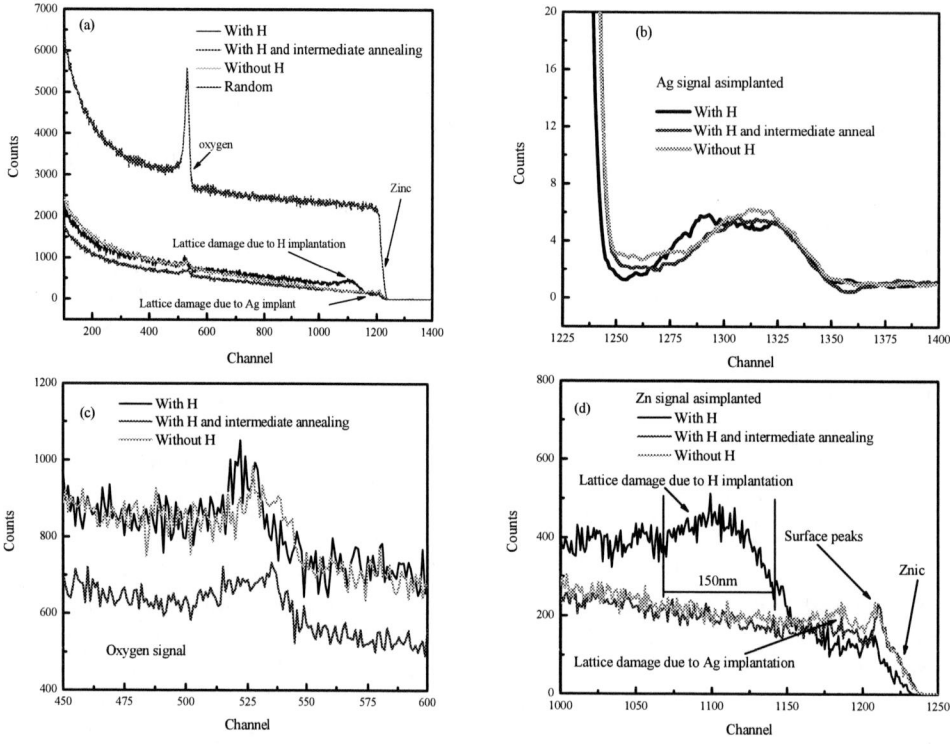

Fig. 1 (a): RBS spectra along <0001> direction for all asimplanted samples with H, without H and intermediately annealed sample. The damage created by the hydrogen implantation can be seen in the bump around 1100 channel number. (b): RBS spectra along <10-11> for Ag signal (c): oxygen signal and (d): zinc signal for all asimplanted samples.

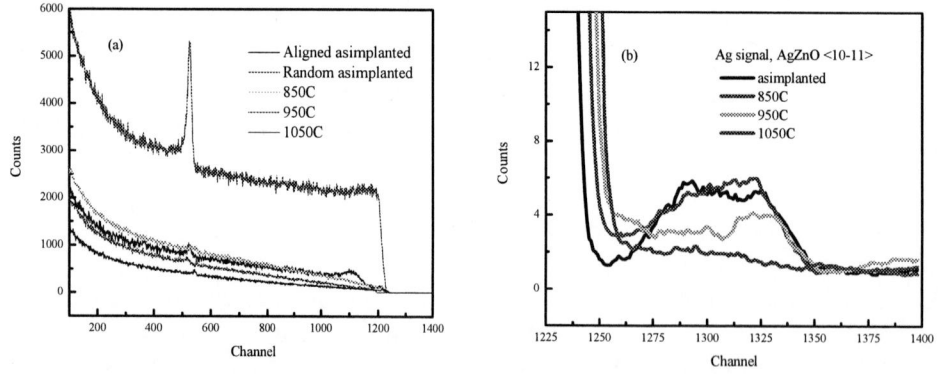

Fig. 2 (a): Asimplanted and annealed samples with H. (b): Ag signal for samples with H.

64

**DISCUSSION**

The $\chi_{min}$ values for zinc and oxygen signals along the <10-11> direction are higher than their values along the <0001> direction because of the better depth resolution in the former case. In case of unimplanted samples, the $\chi_{min}$ values are 3.1 and 3.7% along <0001>, and 22.4 and 12.2% along the <10-11> direction for oxygen and zinc signals respectively. The difference in $\chi_{min}$ values along the two crystal directions is due to the difference in the average scattering potential of the atomic rows. Along the <0001> direction, the Zn and O atoms form two interpenetrating atomic layers in hexagonal closed packed structures and they offer an average potential to the scattering beam, whereas along the <10-11> direction, there are two independent atomic rows of Zn and O atoms. Each atomic row creates its own scattering potential and the analyzing beam is scattered either by Zn or O atomic rows. In Fig. 3 the $\chi_{min}$ values for oxygen is much higher than its values for zinc indicating that the number of displaced oxygen atoms is higher than the zinc displaced atoms after ion implantation. This is expected from the momentum transfer to oxygen atoms during H and Ag implantations. Spratt et al., have seen similar effects in H implanted sapphire crystals [17]. T. Oga, et al., has reported the displaced oxygen atoms as high as twice of the zinc displaced atoms in Al ion implanted ZnO single crystals. [18]. Furthermore, oxygen is very light compare to zinc and probably diffuses to interstitial sites during crystal growth other than its regular lattice sites, resulting in higher $\chi_{min}$ values for O than Zn atoms even in the unimplanted samples. In comparison with the samples without H, the H implanted samples have higher displaced O atoms because of the high chemical reactivity between O atoms and H, which prevents recombination of interstitial O atoms with O vacancies, possibly through the formation of O-H complexes. This theory is further supported by the stronger bond between O and H as compared to the bond strength between Zn and H. Chris Van de Walle has used first principle calculations to show that the H interstitial atoms can bond with O atoms of double strength than they can bond with Zn atoms [19]. This will make the diffusion of O atoms to their regular sites harder resulting in higher number of O atoms between the atomic rows and slower lattice recovery rate for O atoms in H implanted samples. The $\chi_{min}$ of Ag is decreasing with annealing because the Ag atoms are diffusing away from the interstitial positions to Zn substitutional positions. In samples with H the $\chi_{min}$ for Ag at 950°C is higher than the sample without H; this could be possibly due the trapping of Ag atoms in the defects/ cavities created by H implantation. The lower value of Ag $\chi_{min}$ values at 1050°C is due to the annealing out of cavities at higher temperature. More detailed study and structural characterization is needed to understand this effect.

Fig. 3, (d) shows the substitutional Ag atoms on Zn sites. Before annealing the samples, without H implantation samples have the most substitutional Ag atoms and this fraction increases steadily to 61.6% with annealing. The samples with H have initially 26.6% of the Ag atoms at $S_{zn}$, and as the samples are annealed, some of the Ag atoms are diffused into the substitutional sites. At the intermediate temperatures nanocavities are produced and Ag is trapped in resulting in less substitutional Ag atoms. But as the samples are annealed at higher temperature, the cavities are removed and Ag is diffused into the substitutional sites.

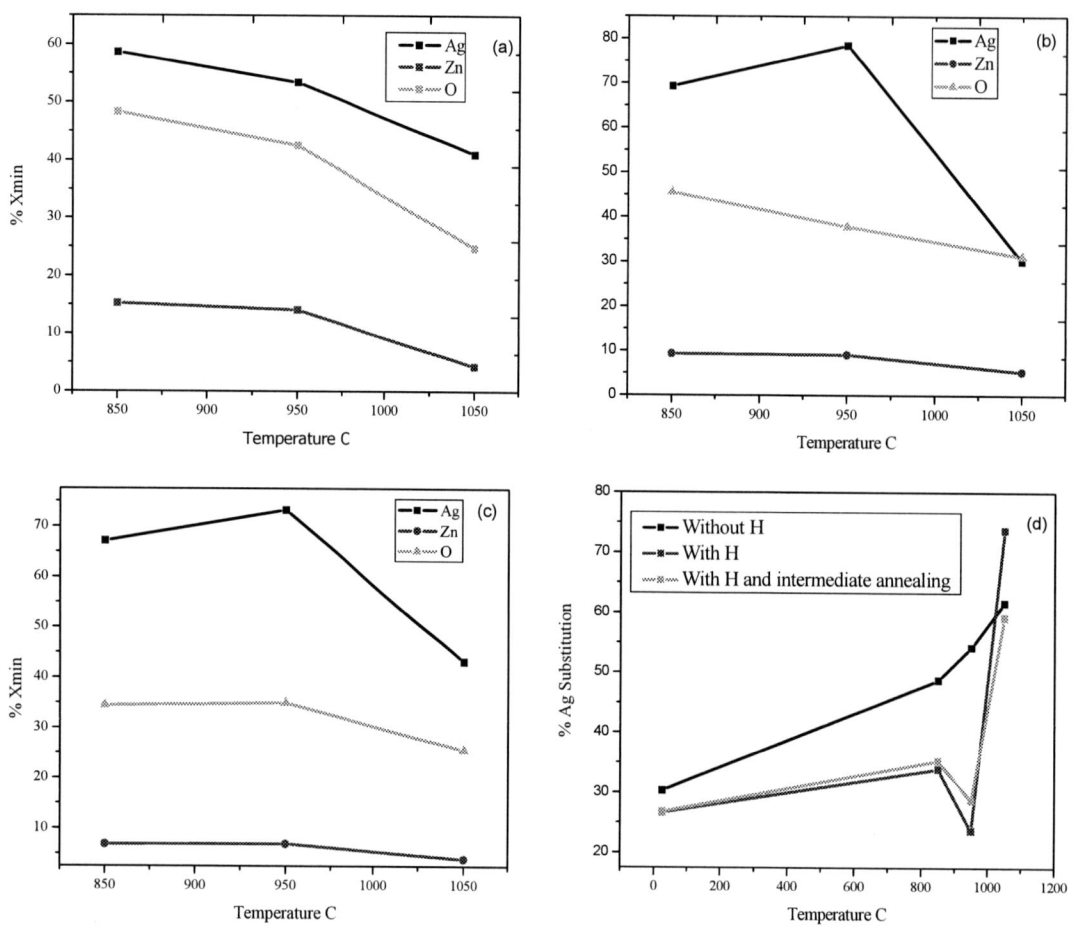

Fig. 3: RBS yield for samples (a); without H, (b); With H, (c); With H and intermediate annealing, (d); Substitution of Ag on Zn sites for all three samples.

## CONCLUSIONS

In conclusions we found that the damage created to oxygen lattice by the Ag atoms during implantation is higher than the damaged caused to the zinc lattice. The Zn lattice is holding stronger than the O lattice during ion implantation. The Zn lattice recovers faster than the O lattice. The Ag substitutional fraction is higher in samples with H at the highest annealing temperature. The H implantation is helping Ag to diffuse into the substitutional sites. Detailed study is needed to understand this phenomenon.

**REFERENCE**

1. D. C. Look, Material Science and Engineering B80, 383, (2001).
2. S. J. Pearton, et al., Superlattices and Microstructures 34, 3-32, (2003).
3. T. Minami, et al. Jpn. J. Appl. Phys., Part 2 **24**, L781 (1985).
4. W. Walukiewicz, Phys. Rev. B **50**, 5221 (1994).
5. C. G. Van de Walle,et al. Phys. Rev. B **47**, 9425 (1993).
6. S. B. Zhang, et al, Phys. Rev. B, 63, 075205, (2001).
7. D. C. Look, et al, Phys. Rev. Lett. **82**, 2552 (1999).
8. D. C. Look, et al. Solid State Commun.**105**, 399 (1998).
9. Hong Seong et al., Applied Physics Letters, **88**, 202108, (2006).
10. E. Rita et al., Hyperfine Interactions, **158**, 395, (2004).
11. Jiwei Fan and Robert Freer, J. Appl. Phys. **9**, 4795, (1995).
12. Isao Sakaguchi, et al., Journal of Ceramic Society of Japan, **118**[3], 217, (2010).
13. Yanfa Yan, et al., Applied Physics Letters, **89**, 181912, (2006).
14. K. Ip, et al., Applied Physics Letters, **81**, 3696, (2002).
15. K. Ip, et al., Applied Physics Letters, **82**, 385, (2003).
16. O. Moutanabbir, et al., Applied Physics Letters, **93**, 031916, (2008).
17. William T. Spratt, et al., Applied physics Letters, **99**, 111909, (2011).
18. T. Oga et al., J. Appl. Phys. **109**, 123702, (2011).
19. Chris. G Van de Walle, Physics Review Letters, **85**, 1012, (2000).
20. Hand Book of Modern Ion Beam Materials Analysis, edited by Joseph. R. Tesmer and Michael Nastasi, Material Research Society, Pittsburgh, Pennsylvania.

Mater. Res. Soc. Symp. Proc. Vol. 1394 © 2012 Materials Research Society
DOI: 10.1557/opl.2012.475

# Influence of post-deposition annealing on structural, optical and electrical characteristics of NiO/ZnO thin film hetero-junction

Manisha Tyagi, Monika Tomar[1] and Vinay Gupta*

Department of Physics and Astrophysics, University of Delhi, Delhi-110007, INDIA

[1]Physics Department, Miranda House, University of Delhi, Delhi-110007, INDIA

*email: vgupta@physics.du.ac.in

## ABSTRACT

Transparent *p-n* hetero-junction diodes are fabricated using, *p*-type NiO and *n*-type ZnO thin films deposited onto a Pt/Ti/glass substrate utilizing RF sputtering technique. The prepared hetero-junctions are studied for the structural, electrical and optical properties and the effect of post-deposition annealing is investigated through *I-V* measurements and XRD analysis. The as deposited hetero-junction is found to be giving ohmic behaviour while with post-annealing treatment it result in rectification with a ratio of forward-to-reverse current as high as 15 in the range -1.0 to 1.0 V. Forward threshold and the reverse breakdown voltages are found to be about 0.5 and -2.7 V, respectively. The forward-bias *I-V* characteristics are dominated by the flow of space-charge-limited current with an optical transmission of above 50 % in the visible region important for the transparent electronic device fabrication.

## INTRODUCTION

The diversity in oxide semiconductor junctions and their functions were rather limited compared to conventional semiconductors, even though oxide semiconductors have excellent stability in harsh environments and have unique functions such as optical transparency, chemical sensing [1], as well as photoluminescence. For exploiting the full potential of oxide semiconductors, a hetero-junction of the *p*-type oxide semiconductor with *n*-type oxide semiconductor has been reported for electronic and optoelectronic device applications [2–6].

ZnO is a promising semiconductor for the fabrication of various optoelectronic devices such as ultraviolet (UV) light emitting diode (LED), transparent conducting electrode in photovoltaics, in display devices and electronic transducers due to the direct wide band gap ($Eg$=3.3 eV) and large exciton binding energy (~60 meV) [7]. However, ZnO is naturally an *n*-type semiconductor and growth of *p*-type ZnO is still controversial. The choice and fabrication of a suitable *p*-type material which could be integrated with *n*-type ZnO for realization of *p-n* hetro-junctions is always an important and crucial issue. Nickel Oxide (NiO) is wide band gap (3.4 to 4.0 eV) semiconducting material showing *p*-type conductivity [8,9]. NiO possess unique optical, electrical and magnetic properties such as high visible transparency, high electrical conductivity along with high carrier concentration and presence of ferromagnetism in nanostructured form. NiO has low lattice mismatch with ZnO, which is beneficial for the fabrication of *p-n* hetero-junction with ZnO [8].

Both ZnO and NiO hold great promise for the fabrication of high temperature and high power optoelectronic devices because of attractive properties these materials possess, such as high breakdown fields, and high thermal conductivities. In addition, electrical conductivity of both NiO and ZnO could be tailored with the change in concentration of native point defects such as

Ni/Zn and O vacancies/interstitials [7,8,10]. To form an electrical junction using ZnO and NiO, especially for minority carrier flow, great care has to be taken to reduce recombination centers.

In this work, a *p-n* hetero-junction diodes based on nickel oxide (NiO)/zinc oxide (ZnO)/corning glass has been fabricated using RF sputtering technique. Effect of post-annealing process on electro-optical performance and *I–V* characteristics of the proposed hetero-structure has been studied. Importance of hetero-junction for the transparent electronic devices has been discussed.

## EXPERIMENTAL DETAILS

For the fabrication of NiO/ZnO *p-n* hetero-junction Pt/Ti coated Corning glass substrate was utilized. A 100 nm thin Pt film was coated by sputtering a Pt (99.99% pure) target in 100% Ar gas ambient at 10 mTorr sputtering pressure to act as bottom electrode. A thin buffer layer of Ti (~20 nm) was deposited prior to Pt deposition to improve its adhesion on corning glass substrate (2 cm × 1 cm). Nickel oxide (NiO) and Zinc Oxide (ZnO) thin films were deposited on Pt/Ti coated Corning glass substrate using a 4 inch diameter metallic nickel (Ni) and zinc (Zn) target (99.99% pure) respectively using RF sputtering technique. The typical deposition parameters are summarized in Table I. For the fabrication of hetero-junctions, 0.5 cm × 0.5 cm area of bottom electrode was masked while depositing the oxide thin films for taking the bottom contact. At first, 110 nm thick ZnO film was deposited followed by deposition of 110 nm thick NiO thin film in-situ for the fabrication of hetero-junctions. NiO and ZnO films were also deposited separately on corning glass substrate under identical conditions for basic characterizations.

**Table I :** Deposition parameters for NiO-ZnO thin film deposition

| Target | Metallic Zinc (99.99%) of 4"dia. | Metallic Nickel (99.99%) of 4"dia. |
|---|---|---|
| Substrate | Masked Pt/Ti/Corning glass | ZnO/Pt/Ti/Corning glass |
| T-S distance | 7.5 cm | 7.5 cm |
| Gas composition | 90 % Ar + 10 % $O_2$ | 100 % $O_2$ |
| Sputtering Pressure | 12 mTorr | 12 mTorr |
| Rf Power | 100 watt | 100 watt |
| Substrate Temperature | Without substrate heating | Without substrate heating |

After deposition, the individual films and hetero-junctions were annealed in a tube furnace at different temperatures (100, 200, 300, 400 and 500 °C) for 2 h in atmospheric air. Pt microdiscs were deposited through a shadow mask (dia.= 600 μm) on the NiO/ZnO/Pt/Ti/glass hetero-junctions for the top electrode. Cross sectional view of the prepared hetero-junctions is shown in figure 1.

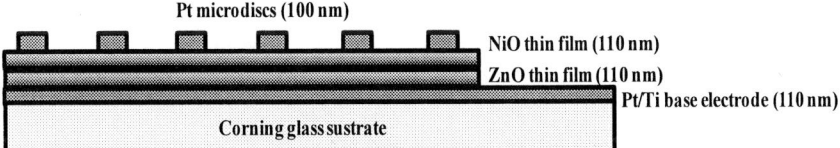

Figure 1. Cross sectional view of Pt/*p*-NiO/*n*-ZnO/Pt/Ti/glass hetero-junction diode.

Structural and optical properties of the were studied using X-Ray diffractometer (Bruker D8 Discover) having Cu K$\alpha_1$ line as monochromatic source ($\lambda$ = 1.54 Å) and a double beam UV-Visible Spectrophotometer (Perkin Elmer, Lambda 35) respectively. Electrical properties of the deposited film including electrical conductivity, carrier concentration and Hall mobility were determined from the Van der Pauw arrangement using Hall measurement system. Thickness measurements of films were carried out using Veeco Dektak 150 A surface profiler. *I-V* characteristics of the devices were obtained using Semiconductor Characterisation System (Keithley 4200).

**RESULT AND DISCUSSION**

**Structural properties**

The as-deposited NiO films were found to be amorphous in nature, which become polycrystalline subject to post-deposition annealing treatment. The film crystallinity is found to be improving with increasing the annealing temperature and is attributed to decrease in native defects [11]. Films annealed at a temperature of 100 °C showed a weak and broadened diffraction peak corresponding to (200) diffraction plane indicating poor crystallinity, while with further increasing the annealing temperature ($\geq$ 200 °C), a preferred oriented sharp diffraction peak corresponding to (200) plane is observed along with diffraction peaks corresponding to weak (111) and (220) planes indicating improved crystallinity of the NiO films. This is due to the fact that annealing provides thermal energy to the atomic species enhancing their mobility. Hence, they have a higher probability to reach the equilibrium position and leads to a crystalline structure. It may be inferred that the NiO films become preferred oriented having most of the crystallites oriented along (200) crystallite plane. On the other hand, ZnO thin films deposited without any substrate heating show only one peak corresponding to the (002) plane which become sharp subject to post-deposition annealing indicating that all films are highly *c*-axis oriented with the (002) plane parallel to the substrate surface. Figure 2 shows the XRD spectra of the NiO/ZnO thin film hetero-junctions deposited at room temperature (curve i) and the one obtained after post-annealing at 300 °C (curve ii) in atmospheric air. It may be seen from the XRD that as-deposited structure shows only one peak corresponding to (002) plane of ZnO and with post-deposition annealing at 300 °C, (200) plane of NiO also appears along with (002) plane of ZnO.

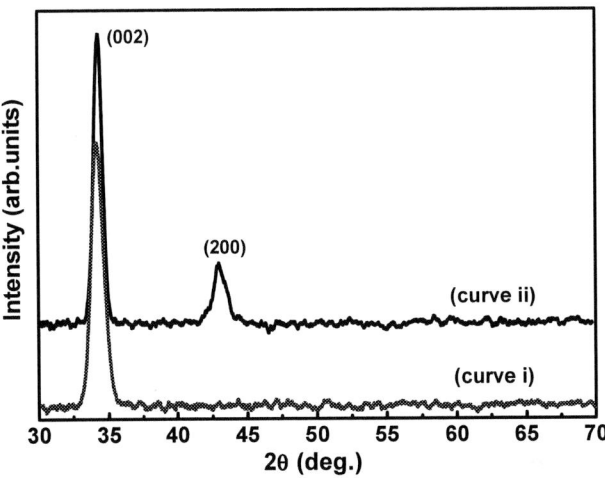

Figure 2. XRD spectra of NiO/ZnO hetero-junctions (curve i): as-deposited and (curve ii): annealed at 300 $^{o}$C in air.

## Optical properties

Nickel oxide and Zinc oxide are the wide (direct) band-gap semiconductors having a well defined absorption edge in the UV region. The optical transmittance of the NiO thin film increases from 18% to 75% with increase in annealing temperature from RT to 500 $^{o}$C whereas there is negligible change in the optical transmission of ZnO thin film after post-deposition annealing. Figure 3 shows the optical transmittance spectra of ZnO and NiO films deposited on corning glass substrate in the range 190-1100 nm after post- annealing at 300 $^{o}$C. The average percentage transmittances of ZnO and NiO thin films are found to be 85% and 65%, respectively. The optical band gaps of ZnO and NiO films as calculated using the Tauc plot are found to be 3.28 and 3.85 eV respectively. The optical transmission of the NiO/ZnO hetero-junctions diode with a total thickness of 220 nm is found to be above 50% in the visible region which is advantageous for the fabrication of transparent electronic devices.

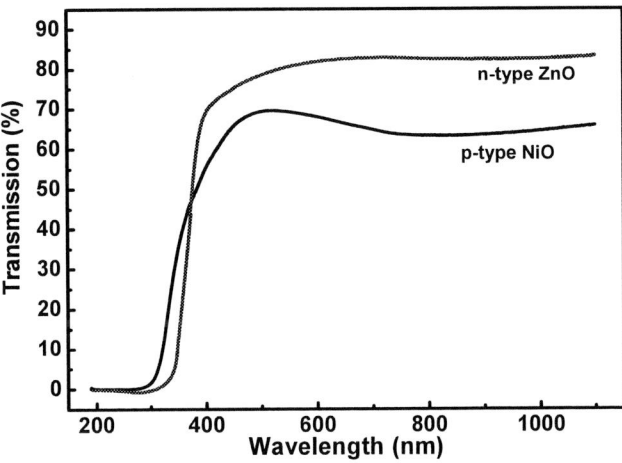

Figure 3. Optical transmission spectra of ZnO and NiO films after post-annealing at 300 $^{o}$C.

The transmission obtained in the present case is much higher compared to the one reported by Sato et al. for their *p*-NiO/*i*-NiO/*i*-ZnO/*n*-ZnO *p-i-n* hetero-junction diode (20% in visible region) and hence shows a potential application of the prepared *p-n* hetero-junctions (NiO/ZnO) device in the transparent electronics.

## Electrical properties

Thermo-emf measurement has been used to evaluate thermoelectric power (TEP) and hence the nature of the majority charge carriers in deposited thin films. Thermo-emf generated in the sample is recorded as a function of temperature of the hot end. All the measurements were recorded at room temperature. The thermally excited majority charge carriers translate within the semiconductor from hot end to the cold end. Thus, the hot end surrounding zone becomes charged with minority carriers and the cold end remains neutral. By this study the nature of charge carriers (whether hole or electron) present in the film can be depicted very easily. Annealed Nickel oxide and Zinc oxide thin films show holes and electrons as the majority charge carriers respectively and hence confirm the *p* and *n*-type nature of the respective films.

Hall measurements show that the as-deposited NiO films possess low resistivity of $4.80 \times 10^{-3}$ $\Omega$-cm. With increase in the post-deposition annealing temperature, the film resistivity is found to be increasing whereas both carrier concentration and mobility of the films are found to be decreasing. The films show positive Hall-coefficient, confirming *p*-type nature of the annealed NiO film and the value of hole concentration is found to be ~$7.33 \times 10^{17}$ cm$^{-3}$ along with the Hall mobility value of 0.60 cm$^{-2}$ V$^{-1}$ S$^{-1}$ for NiO film annealed at 300 °C. Carrier concentration may be related to the defects present in the films, and a decreasing value indicates that the film quality is improving with annealing. Similar results on improvement in the film quality/crystallinity with post-deposition annealing have also been observed by the XRD studies. ZnO thin films annealed at 300 °C possess a resistivity of $1.21 \times 10^2$ $\Omega$-cm with electron carrier concentration of $1.16 \times 10^{15}$ cm$^{-3}$.

Influence of post-deposition annealing on the *I-V* characteristics of the fabricated NiO/ZnO hetero-junction diode has also been studied and is shown in figure 4. The as-deposited films (NiO and ZnO) show a linear behaviour using Pt as contacts, confirming the ohmic nature of these films with Pt electrode. Figure 4 shows the *I-V* characteristics of the as-deposited and annealed hetero-junctions.

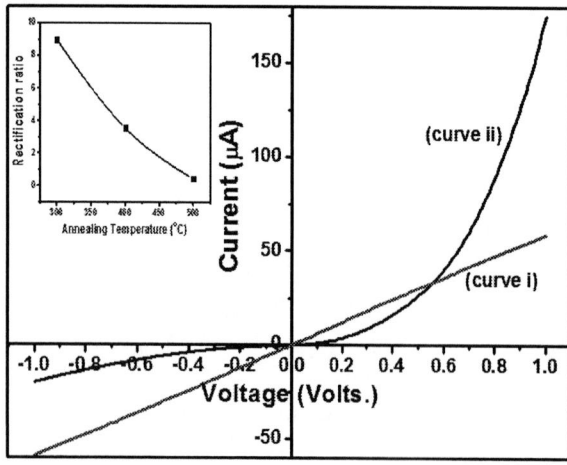

Figure 4. *I-V* characteristics of the as-deposited (curve i) and 300 $^{\circ}$C annealed (curve ii) hetero-junctions.

The as-deposited hetero-junction and the one annealed at 200$^{\circ}$C show an ohmic voltage-current characteristic [curve (i) in figure 4] which is attributed to the tunnelling of carriers through the narrow depletion layer because of the high conductivity of *p*-type layer. With further increasing the annealing temperature to 300$^{\circ}$C (curve (ii) in figure 4) rectifying junction characteristics are observed which degrade with further increasing the annealing temperature (400 and 500 $^{\circ}$C). Rectification ratio obtained for the hetero-junction as a function of post deposition annealing temperature is shown as an inset in figure 4. Three factors may contribute to the changes in *I-V* characteristics subject to post-annealing namely: (a) thermal diffusion of impurities, which results in increasing width of the depletion (space-charge) layer, (b) decrease in the *p*-type conductivity due to thermal oxidation and carrier scattering and (c) increasing series resistance of the *p-n* element [12, 13].The non-linear *I-V* characteristics having diode behaviour could be described by a thermionic emission model. According to the thermionic emission theory, the current in such device could be expressed as [14]

$$I=I_0 \left[\exp(qV/nkT)-1\right] \tag{1}$$

where, $I_0$ is the saturation current, k is the Boltzmann constant, n is the ideality factor, T is the absolute temperature, q is the electron charge, and V is the applied voltage. The barrier height of the device is calculated using the relation [15]

$$I_0 = AA^* \, T^2 \exp(-q\Phi_b/kT) \tag{2}$$

where A is the device area, $A^*$ is the Richardson constant, and $\Phi_b$ is the barrier height. The plot of log I vs. V gives the value of ideality factor. The values of ideality factor and barrier height for the fabricated annealed NiO/ZnO hetero-junctions are estimated to be 3.8 and 0.47 eV, respectively. High value of ideality factor could be attributed to the accelerated recombination of electrons and holes in the depletion region or the presence of interfacial layer [16].

**CONCLUSION**

A transparent hetero-junction based on *p*-NiO and *n*-ZnO has been fabricated using RF sputtering technique successfully. The fabricated junction shows above 50% transmittance in the visible region. Post-deposition annealing at an optimum temperature of 300 $^{\circ}$C is found to be important for obtaining rectifying junction characteristics. A low turn on voltage (0.5 V) and comparatively high transparency in visible region are found to be encouraging for the potential application of hetero-junctions in 'transparent' or 'invisible electronics'.

**REFERENCE**

1. N. Miura, J. Wang, M. Nakatou, P. Elumalai, and M. Hasei, *Electrochem. Solid-State Lett.* **8**, H9 (2005).
2. H. Ohta, K. Kawamura, M. Orita, M. Hirano, N. Sarukura, and H. Hosono, *Appl. Phys. Lett.* **77**, 475 (2000).

3. R. L. Hoffman, B. J. Norris, and J. F. Wager, *Appl. Phys. Lett.* **82**, 733 (2003).

4. X. L. Guo, J. H. Choi, H. Tabata, and T. Kawai, Jpn. *J. Appl. Phys.* **40**, L177 (2001).

5. H. Ohta, M. Hirano, K. Nakahara, H. Maruta, T. Tanabe, M. Kamiya, T. Kamiya, and H. Hosono, *Appl. Phys. Lett.* **83**, 1029 (2003).

6. W. Y. Lee, D. Mauri, and C. H. Wang, *Appl. Phys. Lett.* **72**, 1584 (1998).

7. V. Gupta, A. Mansingh, *Journal of Applied Physics* **80,** 1063 (1996).

8. H. Sato, T. Minami, S. Takata, and T. Yamada, *Thin Solid Films* **236**, 27 (1993).

9. P. Puspharajah, S. Radhakrishna, A.K. Aroif, J. Mater. Sci. 32 (1997) 3001.

10. P.S. Patil, L.D. Kadam, Appl. Surf. Sci. 199 (2002) 211.

11. R.E. Marottia, D.N. Guerraa, C. Bellob, G. Machadoa, E.A. Dalchielea, *Sol. Energy Mater. Sol. Cells* **82** (2004) 85.

12. L. Cattin, B.A. Reguig, A. Khelil, M. Morsli, K. Benchouk, J.C. Bernede, *Appl. Surf. Sci.* **254**, 5814 (2008).

13. M.K. Jayaraj, A.D. Draeseke, J. Tate, R.L. Hoffman, J.F. Wager, Proc. Mat. Res. Soc. Symp. **666 (F4),** 1 (2001).

14. J.M. Shah, Y.L. Li, T. Gessmann, E.F. Schubert, *J. Appl. Phys.* **94**, 2627 (2003).

15. M.M.E. Nahass, K.F.A. Rahman, A.A.A. Darwish, *Microelect. J.* **38,** 91 (2007).

16. R. Singh, A.K. Narula, *Appl. Phys. Lett.* **71,** 2845 (1997).

Mater. Res. Soc. Symp. Proc. Vol. 1394 © 2012 Materials Research Society
DOI: 10.1557/opl.2012.529

# Anomalous Diffusion of Intrinsic Defects in $K^+$ Implanted ZnO using Li as Tracer

L. Vines[1], P.T. Neuvonen[1], A. Yu. Kuznetsov[1], J. Wong-Leung[2,3], C. Jagadish[2] and B.G. Svensson[1]

[1]Department of Physics/Centre for Materials Science and Nanotechnology,
University of Oslo, P.O. Box 1048 Blindern, N-0316 Oslo, Norway
[2]Department of Electronic Materials Engineering, Research School of Physics and Engineering,
The Australian National University, Canberra, ACT 0200, Australia
[3]Centre for Advanced Microscopy, The Australian National University, Canberra, ACT0200,
Australia

## ABSTRACT

Potassium (K) ions have been implanted in hydrothermally grown ZnO to a dose of $1 \times 10^{15}$ cm$^{-2}$, followed by isochronal annealing in a tube furnace (30min) and by rapid thermal annealing (30s) on two separate samples. For annealing temperatures below 700°C, only a minor redistribution of Li is observed behind the projected range of the $K^+$ ions. At temperatures between 700 and 750°C, however, both annealing treatments show a wide region behind the implantation peak which is depleted of Li, and this depletion is used as a tracer to monitor diffusion of intrinsic defects like the Zn interstitial. The results are interpreted as Zn interstitials being released from the implanted region in a burst at temperatures above ~700°C, followed by rapid migration, replacement of Li on Zn site through the kick-out mechanism, and migration of Li away from the active region.

## INTRODUCTION

ZnO is a wide band gap semiconductor exhibiting native n-type conductivity, but the origin of this conductivity is still under debate. On one hand, intrinsic defects like the zinc interstitial ($Zn_I$) and oxygen vacancy ($V_O$), have been reported to be the cause of this native n-type conductivity, but other reports state that the formation energy of $Zn_I$ is too high[1], and the donor level of $V_O$ is too deep[2]. On the other hand, impurities like Al, Ga, In and H are known to be incorporated into ZnO during growth and can act as donors[3-4]. However, none of these impurities alone can explain the donor concentrations observed in experiments[5]. Other important impurities in ZnO are the Group I elements like Li and Na, which may act as either donors or acceptors depending on the atomic configuration. In particular, the concentration of Li is pronounced in hydrothermally (HT) grown ZnO, and it has been shown that Na compete for the same traps as Li, replacing Li by Na in implanted and annealed samples[6]. Moreover, it has recently been found [7] that Li can be used as a tracer element to study the migration of intrinsic elements, in particular $Zn_I$, with dramatic differences in both Li and electrical resistivity profiles after Zn implantation. Specifically, Zn implants result in a Li depletion behind the implantation peak (scaling with dose) after annealing at $\geq 700$ °C, creating a low resistive region (1-10μm) in otherwise highly resistive HT grown ZnO. The Li depletion is not observed when implanting inert elements or elements preferring the O sublattice, and the effect is ascribed to $Zn_I$'s.

In this study, we have implanted Potassium in HT ZnO samples, and measured Li concentration profiles using secondary ion mass spectrometry (SIMS) after annealing in a conventional tube furnace or after rapid thermal annealing (RTA). The SIMS results show that i) the implanted K ions are relatively stable during annealing, i.e. they do not migrate and replace Li, in contrast to Na, ii) excess $Zn_I$'s are formed during implantation resulting in a Li depletion behind the implantation peak, indicating that K prefers the Zn site, and iii) the $Zn_I$ exhibits an anomalous (transient enhanced) diffusion behavior.

## EXPERIMENTAL

A 10x10 mm HT grown ZnO wafer from SPC Goodwill was implanted with 800 keV $K^+$ ions to a dose of $1 \times 10^{15}$ cm$^{-2}$, before it was cut into $5 \times 5$mm$^2$ samples. One sample, no. 1, was subjected to isochronal (30 min) annealing in air ambient in the temperature range of 600 to 1200$^o$C using a conventional tube furnace. A similar study up to 900$^o$C was conducted on sample no. 2, except that the annealing was carried out in a rapid thermal processing chamber, a Jipelec Jetfirst system, with a duration of 30 seconds and a ramp rate of 25$^o$C/s. The impurity concentrations versus depth-profiles were monitored by SIMS using a Cameca IMS7f microanalyzer. 10 keV $O_2^+$ ions were used as a primary beam rastered over a surface area of 150 $\times$ 150 $\mu$m$^2$ and secondary ions were collected from the central part of the craters. Crater depths were measured with a Dektak 8 stylus profilometer, and the erosion rate was assumed to be constant when converting sputtering time to sample depth. Concentration calibration of Li and K were performed using implanted reference samples.

## RESULTS AND DISCUSSION

Figure 1 shows concentration versus depth profiles of K and Li for sample 1 after isochronal annealing (30 min) up to 900$^o$C in tube furnace. A redistribution of K with a plateau around $1 \times 10^{16}$ cm$^{-3}$ and extending $\sim$3$\mu$m towards the bulk is observed after annealing at 700$^o$C, Fig. 1(a), and this redistribution does not change significantly upon further annealing up to 900$^o$C. In fact, a significant migration of K is only observed above 1100$^o$C (not shown). This is a substantially higher temperature than what is observed for the Group I elements Li and Na [6,8], indicating a higher activation energy for K migration. For Li on the other hand, Fig. 1(b), a dramatic redistribution occurs, with a large depleted region after annealing at 700 and 750$^o$C. It has previously been reported that Na implanted and annealed ($\sim$600$^o$C) samples show trap limited diffusion where the migration of Na occurs on the expense of Li [6], indicating that Na and Li compete for the same trap (presumably the zinc vacancy $V_{Zn}$ )[9]. For K, the diffusion is slow and a similar competition with Li is not observed. However, a strong depletion of Li below the implanted region still occurs, even though neither K nor any of the other impurity elements monitored (Al, Mg, Si, Mn, Fe, Ni) change their concentration profiles. A similar effect has been observed for a variety of differently implanted ZnO samples [10-11], and recently it was shown that the effect is particularly pronounced for elements expected to occupy Zn sites, $A_{Zn}$ (hereafter called Zn sublattice elements) [7]. Implanting elements preferring the O sublattice or inert elements, however, do not affect the Li distribution. Accordingly, we conclude that excess $Zn_I$'s are generated when implanted Zn-sublattice elements, and that Li can be used as a trace element to monitor the $Zn_I$ migration. The activation of the $Zn_I + Li_{Zn} \rightarrow Li_I$ reaction occurs in the temperature range 600-800°C.

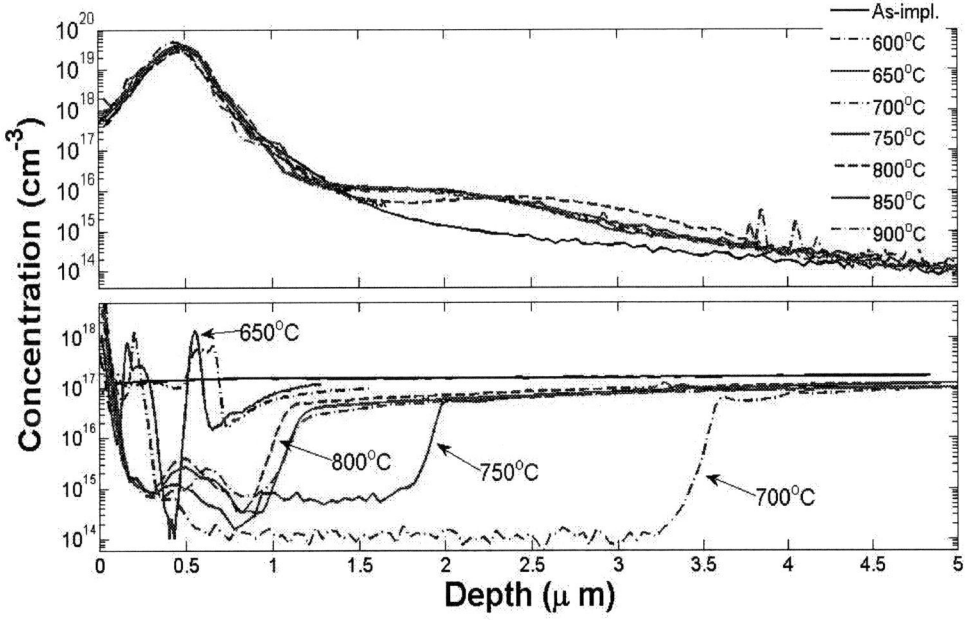

**Figure 1: Concentration versus depth profiles of (a) K and (b) Li after K implantation, with a dose of 1× 10$^{15}$ cm$^{-2}$, and isochronal annealing (30 min) in the temperature range 600 to 900°C**

Furthermore, Fig. 1(b) reveals also that at temperatures above ~800°C a considerable migration of Li occurs from the bulk to the depleted region. This return of Li may suggest that the equilibrium concentration of $V_{Zn}$ is re-established in the depleted region, promoting formation of $Li_{Zn}$ via capturing of migrating interstitial Li ($Li_I$) atoms [12]

Figure 2 shows the profiles of Li and K from sample 2 after RTA for 30 s at temperatures up to 900°C. A shoulder on the K profile is observed in Fig. 2(a) after annealing at 700°C, similar to that in Fig. 1(a), but its extension in depth does not increase at 750°C and the concentration remaining below $\leq 2 \times 10^{16}$ cm$^{-3}$. In Fig 2(b), only a minor redistribution of Li in the implantation peak region occurs at 700°C, but a ~4.5μm deep region below the implantation peak is depleted after the 750°C treatment. Interestingly, this is a larger region than that observed after the furnace annealing at 700 and 750°C (Fig 1). Further increase of the RTA temperature to 800°C only increases the depletion width by ~0.7μm and after 900°C Li migrates back from the bulk, similar to that in Fig. 1. Thus, by comparing the Li profiles after the furnace anneals (30 min) and RTA (30 s) at 700 and 750°C, we can conclude that the apparent diffusivity is time dependent, i.e., a transient diffusion process takes place, resembling that of transient enhanced diffusion (TED) of impurities (dopants) in silicon [13].

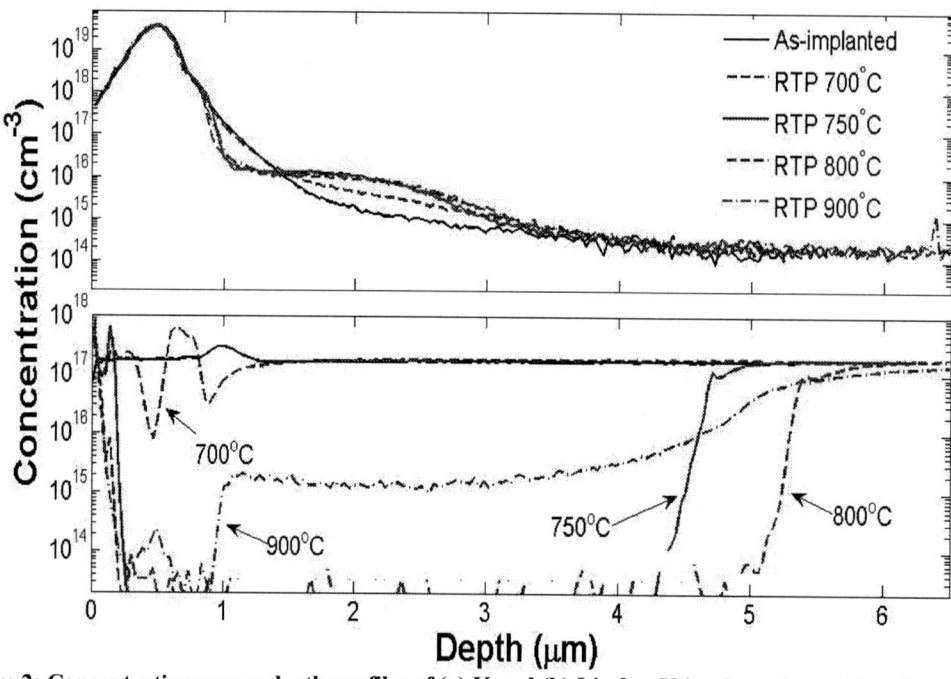

**Figure 2: Concentration versus depth profiles of (a) K and (b) Li after K implantation, with a dose of $1 \times 10^{15}$ cm$^{-2}$, and annealing for 30 seconds in the temperature range 700 to 900°C**

The behavior of Li can be explained by the following scenario. During the K implantation, Zn atoms are displaced, and released as $Zn_I$'s during heat treatment. When a $Zn_I$ encounters $Li_{Zn}$, the lower formation energy of $Zn_{Zn}$ compared to $Li_{Zn}$ results in a kick-out of Li, forming $Li_I$. $Li_I$ subsequently migrates out of the region exposed to $Zn_I$'s, either to the surface or into the bulk. However, the anomalous (transient) diffusion behavior shown in Figs. 1 and 2, indicates that the process is not limited by the migration energy of $Li_I$ or $Zn_I$, i.e. both $Zn_I$ and $Li_I$ migrates rapidly > 700°C. A high diffusivity is in accordance with the activation energies proposed in the literature [14-15]. Thus, the limiting energy barrier for the process, or the anomalous diffusion, is either i) the required energy for the kick-out mechanism or ii) the energy required for releasing $Zn_I$ from the implanted (damaged) region. Scenario (i) assumes a continuous release of $Zn_I$'s from the damaged region, and the depth of the Li depletion should be proportional to the annealing time. However, for a given temperature, e.g. 750°C, the 30 s anneal exhibit deeper Li depletion compared to the 30 min anneal. For (ii), on the other hand, the release of $Zn_I$'s is controlled by a source having a limited concentration, and dependent on the temperature rather than on the duration. For example, Zn related clusters, induced by the implantation, may dissociate, e.g. around 700°C, and release a large amount of $Zn_I$'s, which can migrate and replace $Li_{Zn}$. Hence, the experimental data seem to favor scenario (ii).

The RTA at 900°C results in migration of Li back from the bulk into the previously depleted layer (Fig. 2(b)), except in the implanted region. However, the reappearance of Li occurs at a low concentration, ~ $1 \times 10^{15}$ cm$^{-3}$, indicating that the concentration of available sites for Li is lower compared to that of the bulk (~$2 \times 10^{17}$ cm$^{-3}$). Assuming that the contribution from LiI is small and that the trapping site is mainly $V_{Zn}$, as proposed in [12], it appears that equilibrium concentration of $V_{Zn}$ is not fully established during the 900$^\circ$C treatment in contrast to that for the 900$^\circ$C furnace anneal and also in accordance with the acceptor generation reported for Ga doped ZnO [16].

## CONCLUSIONS

K has been implanted in HT-ZnO to a dose of $1 \times 10^{15}$ cm$^{-2}$, followed by isochronal annealing using regular tube furnace (30min) and rapid thermal processing (30s) on separate samples. For temperatures below 700$^\circ$C, only a minor redistribution of Li is observed below the projected range of the K ions. At annealing temperatures between 700 and 750$^\circ$C, however, both annealing treatments show a wide region ($\leq$5$\mu$m) below the implanted layer depleted of Li. SIMS results show that i) the implanted K is relatively stable during annealing, i.e., it does not migrate and replace Li, in contrast to that of the other Group I elements like Na, ii) $Zn_I$'s are formed during the implantation and released during the subsequent annealing resulting in the depletion of Li below the implanted region and indicating that K prefer the Zn sites, and iii) the $Zn_I$'s exhibit an anomalous (transient) diffusion behavior.

## ACKNOWLEDGMENTS

This work was supported by the Norwegian Research Council through the Frienergi and Nanomat programs and the Australian Research Council through the Discovery projects program.

## REFERENCES

1. A. Janotti and C. G. Van de Walle, *Phys. Rev. B* **76**, 165202 (2007).
2. A. Janotti and C. G. Van de Walle, *Appl. Phys. Lett.* **87**, 122102 (2005).
3. E. V. Monakhov, A. Yu. Kuznetsov, and B. G. Svensson, *J. Phys. D* **42**, 153001 (2009).
4. C. G. Van de Walle, *Phys. Rev. Lett.* **85**, 1012 (2000).
5. A. Janotti and C. G. Van de Walle, *Rep. Prog. Phys.* **72**, 126501 (2009)
6. P.T. Neuvonen, L. Vines, A. Yu. Kuznetsov, B.G. Svensson, X.L. Du, F. Tuomisto, A. Hallen, Interaction between Na and Li in ZnO, *Appl Phys. Lett.*, **95**, 242111, (2009)
7. P.T. Neuvonen, L. Vines, K.E. Knutsen, A. Yu. Azarov, A. Hallèn, B.G. Svensson, A. Yu. Kuznetsov, to be published
8. J.J. Lander, *J. Phys. Chem. Solids*, **15**, 324 (1960)
9. P.T. Neuvonen, L. Vines, V. Venkatachalapathy, A. Zubiaga, F. Tuomisto, A. Hallèn, B.G. Svensson and A. Yu. Kuznetsov, *Phys. Rev. B*, **84**, 205202 (2011)
10. T.M. Børseth, J. S. Christensen, K.Maknys, A. Hall'en, B. G. Svensson, and A. Yu. Kuznetsov, *Superlattices and Microstructures* **38**, 464 (2005).
11. T. Moe Børseth, F. Tuomisto, J. S. Christensen, E. V. Monakhov, B. G. Svensson, and A. Yu. Kuznetsov, *Phys. Rev. B* **77**, 045204 (2008).

12. K. M. Johansen, A. Zubiaga, I. Makkonen, F. Tuomisto, P. T. Neuvonen, K. E. Knutsen, E. V. Monakhov, A. Yu. Kuznetsov, and B. G. Svensson, *Phys. Rev. B* **83**, 245208 (2011).

13. P.A. Stolk, H.J. Gossmann, D.J. Eaglesham, D.C. Jacobson, C.S. Rafferty, G.H. Gilmer, M. Jaraiz, J.M. Poate, H.S. Luftman, T.E. Haynes, *J. Appl. Phys.*, **81**, 6031 (1997)

14. A. Janotti and C.G. Van de Walle, *Phys. Rev. B,* **75**, 165202 (2007).

15. A. Carvalho, A. Alkauskas, A. Pasquarello, A.K. Tagantsev, N. Setter, *Phys. Rev. B*, **80**, 195205 (2009)

16. D.C. Look., K.D. Leedy, L. Vines, B.G. Svensson, A. Zubiaga, F. Tuomisto, D.R. Doutt, L.J. Brillson, *Phys. Rev. B*, **84**, 115202 (2011)

Mater. Res. Soc. Symp. Proc. Vol. 1394 © 2012 Materials Research Society
DOI: 10.1557/opl.2012.530

## Improved Resistive Switching Properties in HfO$_2$-based ReRAMs by Hf/Au Doping

Xiaoli He, Natalya A. Tokranova, Wei Wang and Robert E. Geer
College of Nanoscale Science and Engineering, University at Albany, Albany, NY 12203

## ABSTRACT

Emerging NVM devices have been extensively studied as candidates to extend density scaling and power reduction beyond Si-based flash. Recently, resistive-random-access-memory (ReRAM) devices in the form of metal-insulator-metal (MIM) structures have attracted substantial attention due to their potential scalability, low power operation, and high speed. HfO$_2$ is attractive compared to other transition metal oxides from the vantage point of CMOS process compatibility. Here, we investigate doped HfO$_2$ with a Pt top electrode on an n$^+$-Si substrate. By doping HfO$_2$ with Hf or Au, improved resistive switching properties have been demonstrated in terms of enhanced cycling endurance and lower switching voltages for SET and RESET. The improvements were attributed to doping-induced oxygen vacancies. In addition, Cu-doped HfO$_2$ devices have exhibited multilevel resistive switching.

## INTRODUCTION

Due to the technical and physical limits in highly scaled Si-based flash nonvolatile memories (NVM), emerging NVM devices such as ferroelectric RAM (FeRAM) [1], magneto-resistive RAM (MRAM) [2], phase-change RAM (PCRAM) [3], and recently emerging resistive RAM (ReRAM) [4] have emerged as potential replacements. ReRAM is considered to have promising advantages because of its potential for low power consumption, fast switching and high integration density [4]. Many transition metal oxides including ZrO$_2$ [5], HfO$_2$ [6, 7], and TiO$_2$ [8], have been found to exhibit resistive switching. Different models have been proposed to explain resistive switching, such as formation and rupture of metallic filaments [5, 6], trap-controlled space limited conduction [9], and oxygen migration [8], the universal origins of the resistive switching properties are still a topic of debate. Regardless of the specific switching model, defects and impurities in the oxide layer are expected to play an important role. Previous studies have shown that introducing external defects or traps could improve performance [9]. In this paper, we use HfO$_2$ as the insulator layer for ReRAM devices, due to its CMOS process compatibility, with Pt as the top electrode and n$^+$-Si as the bottom electrode. By doping HfO$_2$ with Hf/Au, improved resistive switching properties have been found in terms of enhanced cycle endurance, lower switching voltages for SET and RESET, and more uniform resistive switching. In addition, multilevel resistive switching of Cu-doped HfO$_2$ devices was demonstrated.

## EXPERIMENT

Resistive switching memory devices (~ 800 μm in diameter) were fabricated on an n$^+$-Si wafer after a chemical cleaning in 5% HF and H$_2$SO$_4$:H$_2$O$_2$ (3:1) solutions. A 30 nm layer of HfO$_2$ was deposited via reactive e-beam evaporation of Hf in an O$_2$ flow. A mini e-beam evaporator system (SPECS EBE-1) with flux stabilization was used for the Hf evaporation. The formation of HfO$_2$ was completed by rapid thermal annealing at 400$^o$C for 1 min. Pt top electrodes were then deposited via conventional e-beam evaporation (Varian 980). All target materials had purity 99.99% and were purchased from Plasmaterials Inc. To promote resistive switching, the $HfO_2$ layer was doped with either Hf or Au. Hf doping was realized by depositing 3 nm of Hf in the absence of $O_2$ flow at the midpoint of $HfO_2$ deposition cycle. For Au doping, $HfO_2$ surfaces were incubated in a solution of citrate-stabilized Au nanoparticles (80 nM, ~Ø5 nm) for 30 min, followed by a rinse and $N_2$ drying. Figure 1 shows the schematic structure of the 3 types of ReRAM devices. Sample morphology and composition were characterized using scanning electron microscopy (SEM, LEO 1550), transmission electron microscopy (TEM, JOEL 2010F) with samples prepared by FIB-SEM (FEI Nova NanoLab 600 Dual Beam), and x-ray photoelectron spectroscopy (XPS, Thermo-VG Scientific XPS system). Electrical measurements utilized a Keithley 4200 characterization system in a DC sweeping mode by applying bias on the top electrode with the bottom electrode grounded.

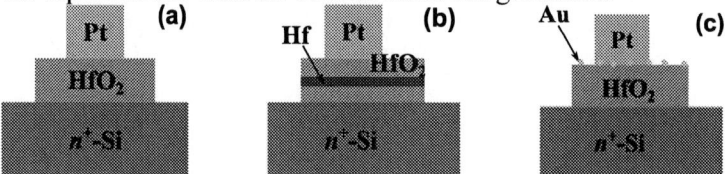

**Figure 1.** Schematic $HfO_2$-based devices: (a) non-doped $HfO_2$, (b) Hf-doped, (c) Au-doped.

## RESULTS AND DISCUSSION

Figure 2a displays a SEM image of the cross-section of a $HfO_2$:Au device. Cross-sections of nondoped $HfO_2$- and $HfO_2$:Hf-based ReRAM devices were similar (not shown). The top layer in Figure 2a is Pt which was deposited for protection from the focused ion beam during cross-section. The measured thickness of the Pt top electrode layer is ~140 nm, and the $HfO_2$ layer is ~27 nm thick. The surface of a non-doped $HfO_2$-based ReRAM device is shown in Figure 2b, where Figure 2b-1 is the surface image on $HfO_2$ and Figure 2b-2 is taken at the Pt top electrode surface. During the experiment, it was found that $HfO_2$:Hf-based ReRAM devices exhibited similar surface morphologies as shown in Figure 2b. Stress-induced cracking of the Pt layer was observed for the device shown in Figure 2b-2. Figure 2c shows the surface images of Au-doped $HfO_2$ ReRAM devices. Figure 2c-1 shows the $HfO_2$ surface and Figure 2c-2 shows the Pt top electrode surface. The micrograph in Figure 2c-1 reveals Au nanoparticles at the $HfO_2$ surface with an average diameter about ~5 nm and density of 625 /$\mu m^2$.

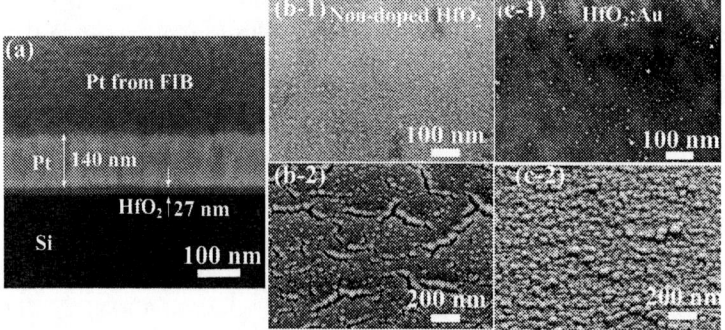

**Figure 2.** SEM images of (a) cross-section of $HfO_2$-based ReRAM devices, (b-1) $HfO_2$ surface, and (b-2) Pt top electrode, for non-doped $HfO_2$-based devices; (c-1) $HfO_2$ surface, and (c-2) Pt top electrode, for Au-doped devices.

The cross-sections of ReRAM devices were investigated with TEM. Figure 3a is taken from a $HfO_2$:Au-based ReRAM device. For that device the Pt top electrode layer thickness is ~147 nm and the $HfO_2$ layer thickness is ~22 nm. Au nanoparticles in this sample were difficult to distinguish at the interface of Pt and $HfO_2$ due to low contrast. A higher magnification image of the Si/$HfO_2$ interface is shown in Figure 3b. An interface oxide layer (~3 nm) is observed between $HfO_2$ and Si. Figure 3c shows a typical cross-section image of $HfO_2$:Hf-based devices. The $HfO_2$ sample clearly has 2 distinct layers, which resulted from the Hf doping.

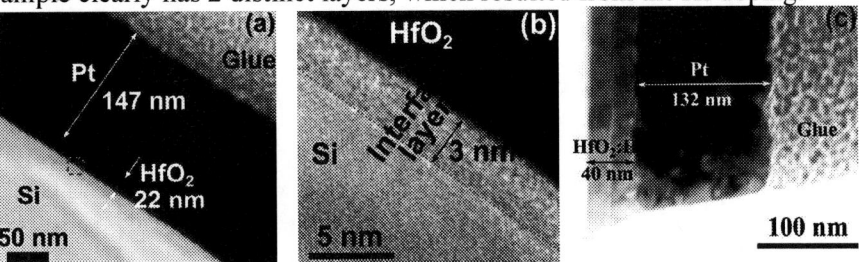

**Figure 3.** TEM images of (a) cross-section of a $HfO_2$:Au-based ReRAM device, (b) higher magnification of the dashed region in (a), (c) cross-section of a $HfO_2$:Hf-based ReRAM device.

The three types of $HfO_2$-based ReRAM devices were further analyzed with XPS. Figures 4a, b and c present the depth profiles for ReRAM devices with non-doped $HfO_2$, Hf-doped $HfO_2$, and Au-doped $HfO_2$, respectively. For non-doped $HfO_2$-based devices (Figure 4a), the relative intensity ratio of Hf and O is constant with the exception of slight increase in O concentration at the interface between $HfO_2$ and Si. This is attributed to the native oxygen occupation on the Si substrate. In Figure 4b, a minimum oxygen concentration is apparent in the middle of the $HfO_2$ region, consonant with the design of the $HfO_2$ layer with Hf-doping. For $HfO_2$:Au-based devices, Au-4f peaks in the XPS spectra appear only at the top level before sputtering (data not shown). This is not unexpected as Au nanoparticles were deposited on the $HfO_2$ surface.

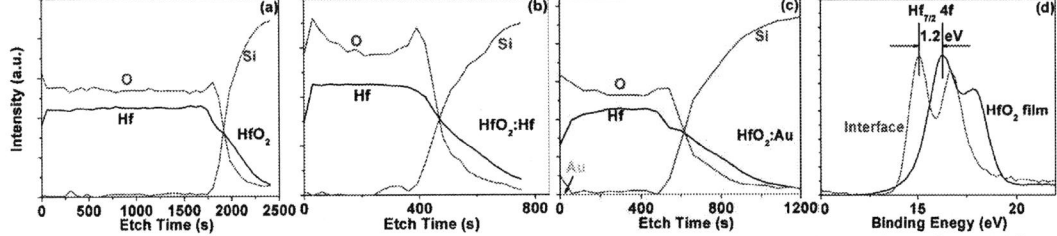

**Figure 4.** XPS depth profiles for (a) non-doped $HfO_2$-based ReRAM devices, (b) $HfO_2$:Hf-based ReRAM devices, (c) $HfO_2$:Au-based ReRAM devices. (d) representative $Hf_{7/2}$ 4f peak shift from 2 different depth levels, each corresponding to $HfO_2$ layer (black) and interface layer (red).

Figure 4d shows the typical $Hf_{7/2}$-4f peaks selected from different sputtering levels (and therefore different depths) that corresponded to the $HfO_2$ film (in black) and the $HfO_2$-Si interface (in red), respectively. The measured $Hf_{7/2}$-4f peak in black color is at the binding energy of 16.24 eV, whereas the red one is at the binding energy of 15.02 eV. Comparing these two $Hf_{7/2}$-4f peaks with the standard binding energy, we conclude that the black trace corresponds to Hf in $HfO_2$ and the red trace corresponds to metallic Hf. This confirms that a distinct Hf-O-Si interface (distinct from a simple $HfO_2$-Si interface) exists in all 3 types of ReRAM devices, consistent with the TEM results (Figure 3b).

Figure 5a shows the typical I-V characteristics of $HfO_2$-based ReRAM devices with non-doped $HfO_2$. The black and red curves are from 2 individual devices (device #1 and #2) respectively. The arrows represent the bias sweeping direction. As seen from the black curve (device #1) in Figure 5a, sweeping a positive bias voltage from 0 on the top electrode, switches the device from its high resistance state (HRS) to its low resistance state (LRS) when the current "jumps" to the preset current compliance (CC). The corresponding set voltage is denoted $V_{set}$. The device in the LRS can be reset to a HRS when applying another positive bias voltage sweep at a voltage denoted as $V_{reset}$. This process is termed unipolar resistive switching since both $V_{set}$ and $V_{reset}$ are of the same polarity. However, bipolar resistive switching for these $HfO_2$-based ReRAM devices is also seen (device #2 in Figure 5a) with $V_{set}$ and $V_{reset}$ at different polarities. Thus, these $HfO_2$-based ReRAM devices have nonpolar resistive switching, as $V_{set}$ and $V_{reset}$ are independent of bias polarity. For devices with non-doped $HfO_2$, the absolute value of set voltage, $|V_{set}|$, is $11 - 15$ V and $|V_{reset}|$ is $7 - 10$ V, which are too large for practical use. Also, these devices have low endurance and no longer exhibit switching after a few set/reset cycles.

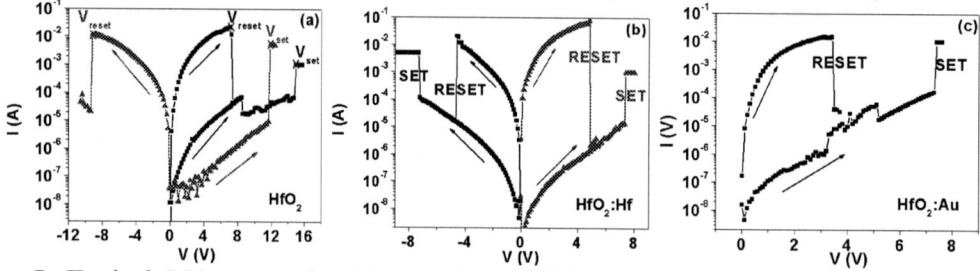

**Figure 5.** Typical I-V curves for (a) non-doped $HfO_2$ ReRAM showing unipolar switching (black - device #1) and bipolar switching (red - device #2), (b) $HfO_2$:Hf ReRAM showing unipolar switching under both polarities (black - device #1, red - device #2), (c) $HfO_2$:Au ReRAM.

The mechanism of resistive switching of these $HfO_2$-based ReRAM devices is generally believed to be intimately related to the migration of oxygen vacancies in $HfO_2$ for conductive filament formation [10]. To enhance oxygen vacancies, a convenient approach is to use Hf-doping to modify the $HfO_2$ stoichiometry. Besides the Hf-doping approach, we also employed Au nanoparticle doping to introduce more defects at the interface between Pt and $HfO_2$ for resistive switching improvement [11].

Typical I-V electrical measurement results of doped $HfO_2$-based ReRAM devices are presented in Figures 5b and c. In Figure 5b for $HfO_2$:Hf devices, the black curve (device #1) shows both set and reset processes under negative bias and the red curve (device #2) shows both set and reset at positive bias. Figure 5c shows unipolar switching at positive bias polarity for $HfO_2$:Au devices. Compared with non-doped $HfO_2$ devices, doped ReRAM devices display lower $|V_{set}|$ and $|V_{reset}|$ values. For $HfO_2$:Hf devices, $|V_{set}|$ ~$7 - 9$ V and $|V_{reset}|$ ~$3 - 5$ V. For $HfO_2$:Au devices, $|V_{set}|$ ~$7 - 10$ V and $|V_{reset}|$ ~$3 - 5$ V. The average values of $|V_{set}|$ and $|V_{reset}|$ for the 3 types of ReRAM devices are summarized in Figure 6a, indicating improvement with respect to lower $|V_{set}|$ and $|V_{reset}|$ and better uniformity with smaller deviations. Also, the ReRAM devices with doping exhibited enhanced endurance (> 30 cycles) (not shown here). Thus, $HfO_2$-based ReRAM devices doped with Hf and Au show improved performance. Resistances of the devices at HRS and LRS are read at $V_{read}$ = +/- 1 V. $R_{off}$ of all ReRAM devices ranged from $10^6$ – $10^{10}$ Ω. However, as seen in Figure 6b, the average $R_{on}$ has smaller values in devices with Hf-doping (3800 Ω) and Au-doping (560 Ω) compared to an average $R_{on}$ of ~22 kΩ for non-doped devices.

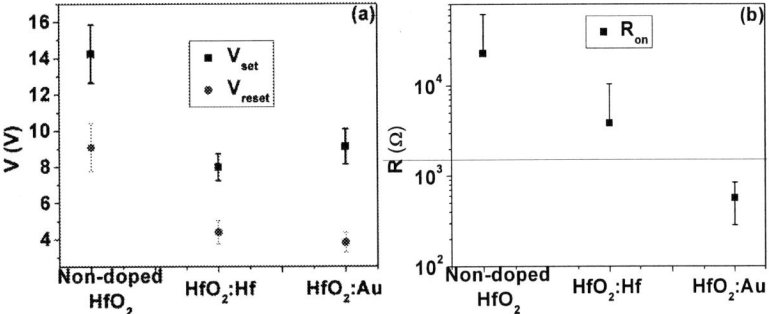

**Figure 6.** Comparison of parameters for HfO$_2$-based ReRAM devices; (a) average $V_{set}$ and $V_{reset}$, (b) average $R_{on}$. $R_{off}$ exhibited the same range ($10^6$-$10^8$ Ω) for all three types of devices.

Large $V_{set}$ and $V_{reset}$ were observed in HfO$_2$-based ReRAM devices showed above. This may be due to the interface layer between HfO$_2$ and Si. Consequently, we modified the structure using Pt as the bottom electrode on a Si/SiO$_2$/Ti substrate and Cu as the top electrode with Cu-doped HfO$_2$. The schematic structure for this stack is shown in the inset of Figure 7a. Repetitive I-V measurements show that resistive switching for this type of ReRAM device is bipolar, that is, the set process takes place under positive bias and the reset process under negative bias with $V_{set}$ ~0.4 – 0.8 V and $V_{reset}$ ~ -0.4 – - 0.8 V. During the measurement with CC ~ 1mA, multiple discrete resistance-change steps appear in both set and reset processes for each set/reset cycle, as shown in the dashed region in Figure 7a. Figure 7b shows a zoomed-in view of the dashed region for the set process shown in Figure 7a. The dashed region for the reset process is not shown here since the resistance steps are similar. Five discrete resistance steps with different values ranging from 20 kΩ to 450 Ω were noted. After increasing the current compliance to 10 mA, there were three major resistance steps, as shown in Figure 7c: R1 ~40 kΩ; R2 ~700 Ω; and R3 ~30 Ω. A similar multi-step switching behavior was demonstrated in other ReRAM devices including devices based on Cu/ZrO$_2$:Cu/Pt and Ag/Ag-Ge-S/W stacks [12, 13] and may be related to the formation and rupture of multiple conducting filaments [14].

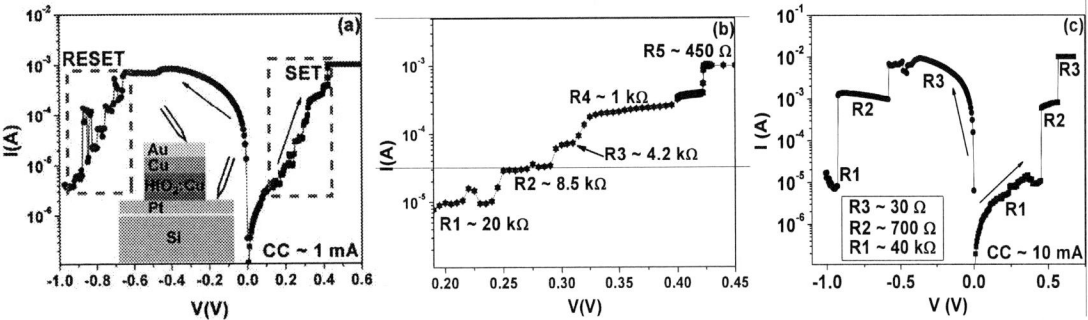

**Figure 7.** (a) Typical I-V curve for Cu/HfO$_2$:Cu/Pt ReRAM devices with multiple resistance steps (1mA CC). (b) Finer I-V plot of dashed region in set process. (c) I-V curve for Cu/HfO$_2$:Cu/Pt ReRAM with multiple resistance steps in both set and reset process (10 mA CC).

CONCLUSIONS

Pt/HfO$_2$/n+-Si ReRAM devices have been fabricated and doped with Hf or Au nanoparticles. The resistive switching behavior was compared, demonstrating that HfO$_2$-based ReRAM devices have nonpolar switching behavior and HfO$_2$-based ReRAM devices with Hf or Au doping show improved resistive switching performance in terms of lower $V_{set}$ and $V_{reset}$, better stability, and cycle endurance. Moreover, with Hf or Au-doping, the HfO$_2$-based ReRAM devices show larger conductivity at LRS. The improvement in resistive switching can be attributed to enhanced defects and oxygen vacancies inside HfO$_2$. The large $V_{set}$ and $V_{reset}$ ($|V_{set/reset}| > 3$ V) might be related to the interfacial layer between Si and HfO$_2$. Using Pt as the bottom electrode and Cu as the top electrode, bipolar resistive switching was observed with low $V_{set}$ and $V_{reset}$ ($|V_{set/reset}|<1$V). With this type of HfO$_2$-based ReRAMs, multiple resistance steps were observed during set/reset process and therefore multi-level resistive switching can potentially be achieved. HfO$_2$-based ReRAM devices with doping may constitute an attractive technology route for non-volatile memory applications.

## ACKNOWLEDGMENTS

This work is partially supported by NSF (grant No. 0829824), SRC FCRP through IFC, International SEMATECH, AFRL (grant No. FA8750-10-1-0138) and IBM research grants.

## REFERENCES

1. H. Shiga et al, IEEE J. Solid-state Circuits **45**, 142 (2010).
2. Y. Huai, AAPPS Bulletin **18**, 33(2008).
3. G. W. Burr et al, J. Vac. Sci. Technol. B **28**, 223 (2010).
4. R. Waser, R. Dittmann, G. Staikov and K. Szot, Adv. Mater. **21**, 2632 (2009).
5. M. Liu, Z. Abid, W. Wang, X. He, Q. Liu and W. Guan, Appl. Phys. Lett. **94**, 233106 (2009).
6. B. Butcher, X. He, M. Huang, Y. Wang, Q. Liu, H. Lv, M. Liu and W. Wang, Nanotechnol. **21**, 475206 (2010).
7. Y. Wang et al, Nanotechnol. **21**, 045202 (2010).
8. J. J. Yang, M. D. Pickett, X. Li, D. A. A. Ohlberg, D. R. Stewart and R. S. Williams, Nature Nanotechnol. **3**, 429 (2008).
9. Q. Liu, W. Guan, S. Long, M. Liu, S. Zhang, Q. Wang and J. Chen, J. Appl. Phys. **104**, 114524 (2008).
10. M. Y. Chan, T. Zhang, V. Ho and P. S. Lee, Microelectronic Engineering **85**, 2420 (2008).
11. L. Goux, X. P. Wang, L. Pantisano, N. Jossart, B. Govoreanu, J. A. Kittl, M. Jurczak, L. Altimine and D. J. Wouters, Electrochem. Solid-State Lett. **14**, H244 (2011).
12. Q. Liu, C. Dou, Y. Wang, S. Long, W. Wang, M. Liu, M. Zhang and J. Chen, Appl. Phys. Lett. **95**, 023501 (2009)
13. U. Russo, D. Kamalanathan, D. Ielmini, A. L. Lacaita, and M. N. Kozicki, IEEE Trans. Electron Device, **56**, 1040 (2009).
14. H. Manem, G. S. Rose, X. He and W. Wang, in *Proc. of the 20th ACM Great Lakes Symposium on VLSI 2010* (Providence, RI, May 2010), pp. 287-192.

Mater. Res. Soc. Symp. Proc. Vol. 1394 © 2012 Materials Research Society
DOI: 10.1557/opl.2012.531

# Heteroepitaxial growth of ZnO films on $Gd_3Ga_5O_{12}$ garnet substrates

Yosuke Ono[1], Hiroaki Matsui[1,2], and Hitoshi Tabata[1,2]

[1]Department of Electrical Engineering and Information Systems, the University of Tokyo, Tokyo 113-0032, Japan

[2]Department of Bioengineering, the University of Tokyo, Tokyo 113-0032

## ABSTRACT

This study focused on structural and optical properties of ZnO films grown epitaxially on $Gd_3Ga_5O_{12}$ substrates. ZnO films ($a$ = 3.2439 Å and $c$ = 5.2036 Å) were deposited on the (001) and (111) planes of $Gd_3Ga_5O_{12}$ (GGG: $a$ = 12.383 Å) garnet substrates by a pulsed laser deposition method. From out-of-plane and in-plane X-ray diffraction measurements, the obtained ZnO films showed a single phase with the (0001) orientation on the GGG (001) and (111) substrates. The epitaxial relations between the ZnO film and GGG (001) substrate were [10-10] ZnO ‖ [100] GGG and [10-10] ZnO ‖ [010] GGG, while the epitaxial relations between the ZnO film and GGG (111) substrate were [10-10] ZnO ‖ [11-2] GGG ±21°. Furthermore, transmittance electron microscopy revealed sharp interfaces between ZnO films and GGG substrates. From photoluminescent spectra, the ZnO films showed donor bound emissions superimposed with free excitons at a low temperature of 10 K.

## INTRODUCTION

Rare earth (RE) garnets, $Re_3Fe_5O_{12}$, have received much attention because of their magnetism and magneto-optical properties. These properties are expected to be promising for practical applications such as scintillations [1], microwave magnets [2], and optical isolators [3]. Epitaxial layer growth of garnets is usually performed using garnet-types of substrate in order to obtain high quality garnet films. However, $Re_3Fe_5O_{12}$ single crystals with micrometer sizes are markedly limited for device applications because it is not easy to use garnet crystals with micrometer sizes. Unfortunately, it has not been achieved epitaxial layer growth of garnet materials on non-garnet substrates. To date, heteroepitaxy on garnet substrates has been only reported using compound semiconductors such as InP [4] and GaAs [5]. In this work, we found that ZnO grew epitaxially on the GGG (001) and (111) substrates. In order to achieve epitaxial layer growth of garnet materials on non-garnet substrates, it is indispensable for understanding epitaxial layer growth of non-garnet materials on garnet substrates as a first step.

In this paper, we report the higher ordered epitaxial growth of ZnO on $Gd_3Ga_5O_{12}$ (GGG) substrates. This growth technique is applied for ZnO epitaxial layers on $c$-face $Al_2O_3$ substrates [6]. We believe that this study contributes to elucidate origin of heteroepitaxial growth of non-garnet materials on garnet substrates.

## EXPERIMENT

Prior to film growth, GGG (001) and (111) substrates were annealed at a high temperature of 1250 °C for 2.5 hours. ZnO films were grown on the GGG substrates at 800 °C by pulsed laser ablation. ArF excimer laser pulses (193 nm, 3Hz and 4 J/cm$^2$) were focused on ZnO ceramic targets (5N) away 5 cm from the substrates in oxygen ambient of 1.8 x 10$^{-1}$ Pa. Thicknesses of the films on GGG (001) and (111) substrates were 98 and 96 nm, respectively.

Atomic force microscopy (AFM) was used to observe the surface morphologies of $Gd_3Ga_5O_{12}$ substrates and ZnO films. Crystal quality of the films was investigated by high-resolution x-ray diffraction (HR-XRD). Structural observations at nano scale were performed by transmission electron microscopy (TEM). Absorption spectra of the films were measured by a UV-visible spectrometer at temperatures of 30, 150 and 300 K. Photoluminescence (PL) spectroscopy was investigated using a He-Cd laser (325 nm) excitation source with light power density of 5 mW and a 0.5-m single monochromator equipped with a 1200 grooves / cm grating blazed at 500 nm.

## DISCUSSION

AFM images of top surfaces on the substrates and films are shown in Fig. 1. Step and terrace structures at atomic scale were clearly observed on both GGG substrates (Fig. 1(a) and 1(b)). The ZnO films showed rough surfaces with three-dimensional (3D) growth. In particular, hexagonal facetted form was observed in the ZnO films grown on GGG (111) substrates, which have been seen on the ZnO films with (-c) polarity grown at high temperatures on $Al_2O_3$ (0001) substrates [7].

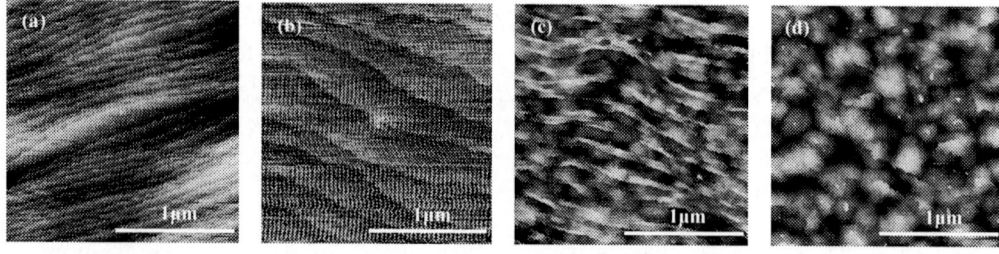

**Figure 1**. AFM images of (a) GGG (001), (b) GGG (111), (c) ZnO/GGG (001), and (d) ZnO/GGG (111).

XRD was employed to investigate the crystal quality of ZnO films on GGG (001) and (111) (Fig. 2). Diffraction peaks at 21.8°, 44.3° and 64.8° were ascribed to the substrates and a measurement stage of XRD. $2\theta\text{-}\omega$ scans showed a series of the ZnO (0002) peaks. These peaks indicated that the ZnO films on GGG (001) and (111) had $c$-axis orientations. In the first case, the $2\theta\text{-}\omega$ scan showed the ZnO (0002) peak at $2\theta = 34.42°$. A lattice constant along the $c$-axis of the ZnO film was calculated to be 5.212 Å, which indicated that $c$-axis length were 0.16% larger than that of a ZnO crystal ($c = 5.204$ Å). While in the second case, the $2\theta\text{-}\omega$ scan showed the ZnO (0002) peak at $2\theta = 34.50°$. A lattice constant along the $c$-axis of the ZnO film was calculated to be 5.200 Å, which was close to that of a ZnO crystal.

The in-plane alignments of the ZnO films grown on the GGG (001) and (111) substrates were measured using the $\phi$ scan of the HR-XRD (Fig. 3). The (10-11) plane of the ZnO films showed twelve symmetries on the GGG (001) and (111) substrates. These twelve peaks indicated that both of the ZnO films were epitaxially grown on GGG substrates with double domains. The (224) planes of the GGG (001) and (111) substrates with four- and three-fold symmetry were also measured for comparison, respectively. In the case of the ZnO films grown on GGG (001), the twelve peaks in the $\phi$ scan corresponded to six peaks originated from ZnO film had two different in-plane orientations, which were rotated 30° around the surface normal from each other (Fig. 3(a)). The in-plane alignments of the ZnO films grown on the GGG (001) were similar to that of the ZnO films grown on $Al_2O_3$ (0001) [8]. The in-plane epitaxial relations between the ZnO film and GGG (001) substrate were [10-10] ZnO || [100] GGG and [10-10] ZnO || [010] GGG. While in the case of ZnO grown on GGG (111), the twelve peaks in the $\phi$ scan corresponded to six peaks originated from ZnO film also had two different in-plane orientations.

However, rotation angle of two domains was 42°. These two domains were asymmetrically rotated with rotation angles of ±21° with respect to [11-2] GGG direction from [10-11] ZnO direction. This asymmetry is called the mirror symmetry [9]. The mirror symmetry is observed when the main symmetry directions of two domains misalign. The in-plane epitaxial relations between the ZnO film and GGG (111) substrate were [10-10] ZnO || [11-2] GGG ±21°. Also, sizes of diffraction spots were smaller than that of ZnO film grown on GGG (001). These indicated that it was a little bit of the stretch for ZnO to grow on GGG (001).

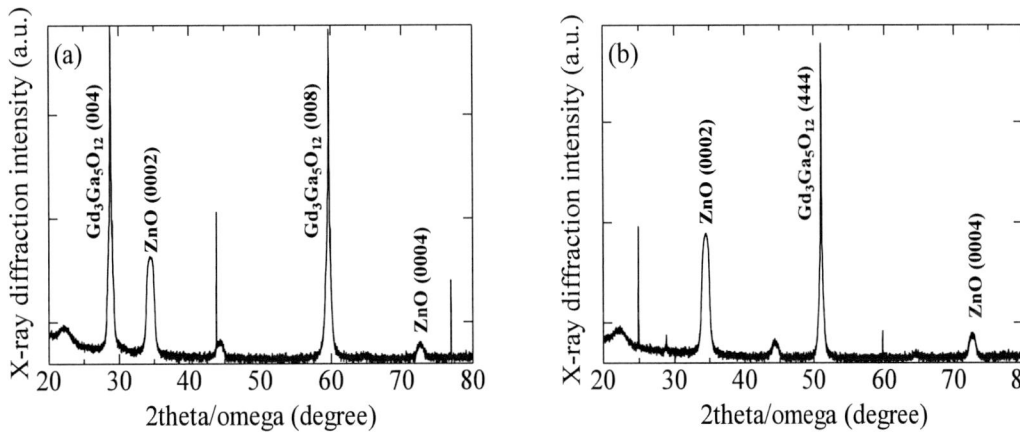

**Figure 2**. (a) $2\theta$-$\omega$ patterns of ZnO films grown on $Gd_3Ga_5O_{12}$ substrates. (b) $2\theta$-$\omega$ patterns of ZnO films grown on $Gd_3Ga_5O_{12}$ (111) substrates. Diffraction peaks at 21.8°, 44.3° and 64.8° are ascribed to the substrates and a measurement stage of XRD.

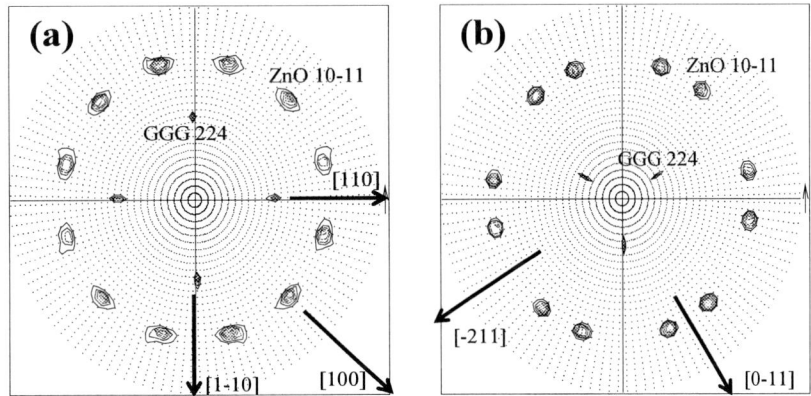

**Figure 3**. X-ray pole figures for the (10-11) plane of ZnO films grown on (a) $Gd_3Ga_5O_{12}$ (001) and (b) $Gd_3Ga_5O_{12}$ (111) substrates.

The cross-sectional TEM measurements of the ZnO films grown on GGG (001) and (111) substrates were performed (Fig. 4). The zone axis for GGG substrate was identified as [110] GGG. In the first case, many treading dislocations were observed (Fig. 4(a)). While in the second case, a few treading dislocations were observed (Fig. 4(b)). These results agreed with those of X-ray pole figure measurements. Also, the HR-TEM images of the interfaces between the ZnO films and GGG substrates were taken in the [110] zone axis of GGG substrates (Fig. 4(c) and (d)). Atomically sharp interfaces and no buffer layer were seen at the interfaces between the ZnO films and GGG substrates. The selected area electron diffraction patterns (Fig. 4(c) and (d) insets) confirmed [0001] ZnO ∥ [001] GGG and [0001] ZnO ∥ [111] GGG orientations in agreement with XRD observations.

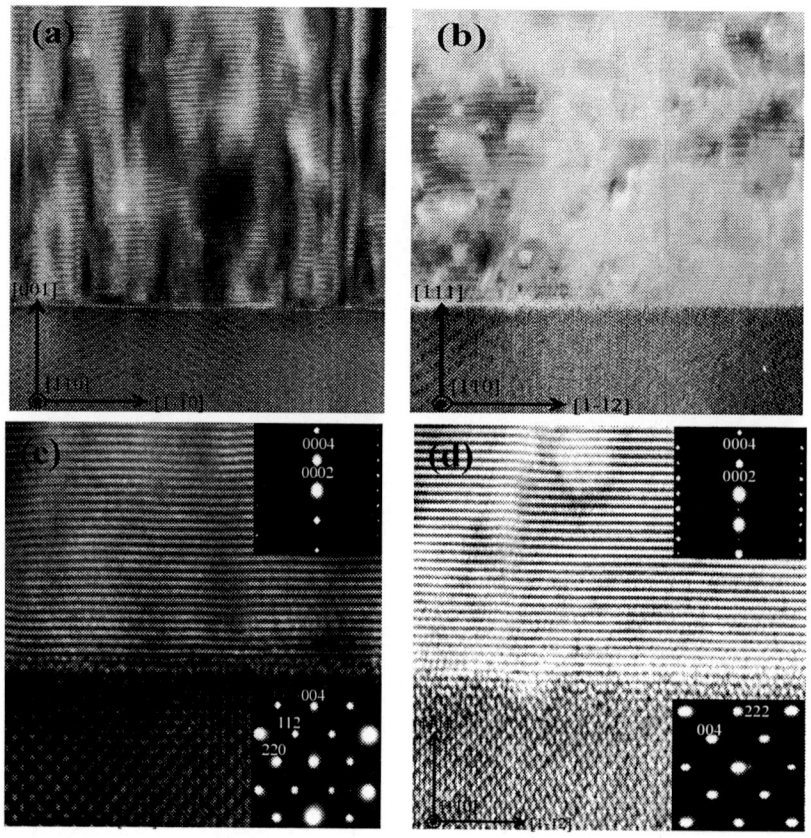

**Figure 4**. TEM images of the ZnO films grown on (a) GGG (001) and (b) GGG (111) substrates. HR-TEM images of the ZnO films grown on (c) GGG (001) and (d) GGG (111).

Absorption spectra of ZnO films on $Gd_3Ga_5O_{12}$ (001) and (111) substrates at different temperatures were measured (Fig. 5(a) and (b)). At 30 K, a free-excitonic transition (FX) was clearly observed at 3.38 eV with phonon side bands of 72 meV. The FX peak gradually shifted to low energy region with increasing temperature based on a narrowing of band gap. The temperature dependence of absorption on the FX peak for ZnO film on $Gd_3Ga_5O_{12}$ (001) substrate was similar to that for ZnO film on $Gd_3Ga_5O_{12}$ (111) substrate.

PL spectra of ZnO films grown on $Gd_3Ga_5O_{12}$ (001) substrates were shown in Fig. 6 (a). From FX peaks in absorption spectra, PL peaks related to bound excitons ($D^oX$) were mainly observed in the temperature range of 10-300 K. PL intensities of $D^oX$ peaks gradually decreased with an increase of temperature. In contrast, PL spectra of ZnO films on $Gd_3Ga_5O_{12}$ (111) substrates exhibited a similar temperature dependence (Fig. 6(b)).

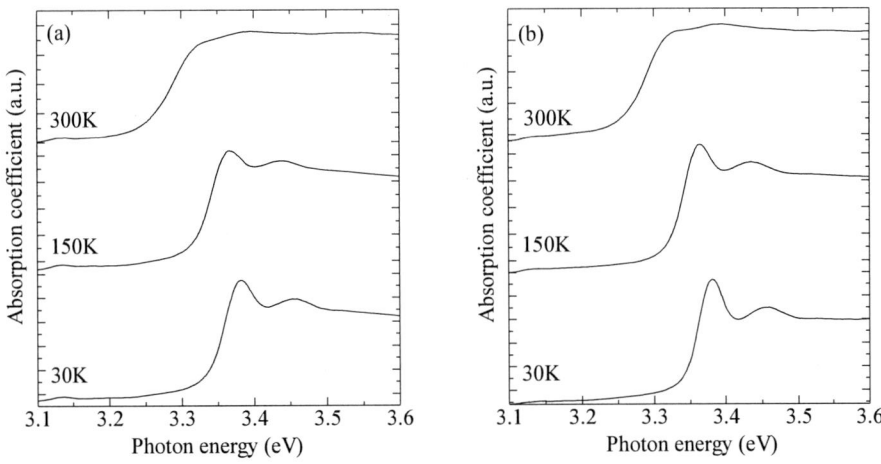

**Figure 5**. Absorption spectra of ZnO films grown on (a) GGG(001) and (b) GGG(111) substrates.

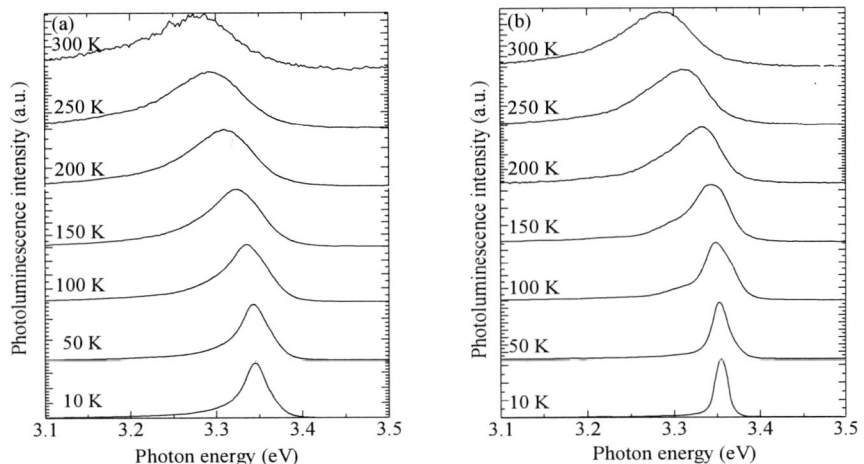

**Figure 6**. PL spectra of ZnO films grown on (a) GGG(001) and (b) GGG(111) substrates.

## CONCLUSIONS

This study for the first time showed heteroepitaxial growth of ZnO films on the (001) and (111) orientations of GGG substrates with double domains. The out-of-plane and in-plane epitaxial relations between ZnO film and GGG (001) substrate were [0001] ZnO ∥ [001] GGG and [10-10] ZnO ∥ [100] [010] GGG. The out-of-plane and in-plane epitaxial relations between ZnO film and GGG (111) substrate were [0001] ZnO ∥ [001] GGG and [10-10] ZnO ∥ [11-2] GGG±21°. As a further challenging task, we are aiming at achieving heteroepitaxial growth of garnets on non-garnet materials. We believe that this study will be the first step of achieving this goal.

## REFERENCES

1. Nerine J. Cherepy, Joshua D. Kuntz, Thomas M. Tillotson, Derrick T. Speaks, Stephen A. Payne, B. H. T. Chai, Yetta Porter-Chapmanm, Stephen E. Derenzo, nuclear Instruments and Methods in Physics Research A 579 (2007) 38
2. Vincent G. Harris, Handbook of Magnetic Materials 20 (2012) 1
3. Hanju Lee, Taedong Kim, Songhui Kim, Youngwoon Yoon, Seungwhan Kim, Arsen Babajanyan, Takayuki Ishibashi, Barry Friedman, and Kiejin Lee, Journal of Magnetism and Magnetic Materials 322 (2010) 2722
4. J. Haisma, A.M.W. Cox, B.H. Koek, D.Mateika, J.A. Pistorius, and E.T.J.M. Smeets, Journal of Crystal Growth 87 (1988) 180
5. J. Haisma, B.H. Koek, J.W.F. Maes, D. Mateika, J.A. Pistorius, and P.J. Roksnoer, Journal of Crystal Growth 83 (1987) 466
6. T. Ohnishi, A. Ohtomo, I. Ohkuboa, M. Kawasaki, M. Yoshimoto, and H. Koinuma, Materials Science and Engineering B56 (1998) 256
7. Z. K. Tang, G. K L. Wong, P. Yu, M. Kawasaki, A. Ohtomo, H. Koinuma, and Y. Segawa, Applied Physics Letters 72 (1998) 25
8. J. M. Yuk, Y. S. No, T. W. Kim, J. Y. Kim, and W. K. Choi, Journal of the Korean Physical Society 55 (2009) 246
9. Marius Gundmann, Physica Status Solid B 248 No.4 (2011) 805

Mater. Res. Soc. Symp. Proc. Vol. 1394 © 2012 Materials Research Society
DOI: 10.1557/opl.2012.654

## ZnO Coated Nanoparticle Phosphors

Masakazu Kobayashi [1,2]

[1]Department of Electrical Engineering and Bioscience, Waseda University, 3-4-1 Okubo
Shinjuku Tokyo, 169-8555 Japan
[2]Kagami Memorial Laboratory for Materials Science and Technology, Waseda University, 2-8-
26 Nishiwaseda, Shinjuku Tokyo, 169-0051 Japan

## ABSTRACT

Conventional phosphor materials are doped ternary or quaternary compounds; hence it would be difficult to prepare nanoparticles of those materials by build up methods. $Ba_2ZnS_3$:Mn (BZS), $SrGa_2S_4$:Eu, and $BaAl_2S_4$:Eu nanoparticles were prepared by a break down method, namely the ball-milling method. Transmission electron microscopy (TEM) and TEM- energy-dispersive X-ray spectroscopy (EDX) measurements showed several-nanometer-size stoichiometric and dispersed nanoparticles were achieved. ZnO-coating was performed and the uniform coating layers were formed on the phosphor nanoparticles. The ZnO-coated nanoparticles exhibited an improved stability in Photoluminescence. Red color phosphor material, namely BZS, was ball-milled and sprayed on the glass substrate. Mn doped BZS absorbs ultra violet light and emits red light peaking at around 640nm. When the single crystal Si solar cell was placed under the transparent nanoparticle layer, short wavelength light was absorbed and converted to long wavelength light.

## INTRODUCTION

Electroluminescence devices are widely expected to have the high brightness and high color purity because of the self light-emission. II-VI compounds such as ZnS and SrS, and thiogalate sulfides such as $(Ca,Sr)Ga_2S_4$ have been studied as qualified phosphors, where some rare-earth-ions are introduced as emission centers. There are two types of the light emission mechanism in nanoparticle phosphor materials. In the case of CdSe/ZnS [1-3] and CdSe/CdS [4,5] nanoparticles, the recombination of electron and hole are used for the light emission. On the other hand, $ZnS:Mn^{2+}$ nanoparticles [6], use an inner shell transition to obtain the light emission. We have prepared CdSe and $ZnS:Mn^{2+}$ nanoparticles by a break down method, namely the ball-milling method. Nanoparticles with good optical properties were achieved [7,8]. The ball milling method is a simple method and would be suited to mass production. Since Cd and Se are harmful elements to the human body and CdSe nanoparticles require rigorous control of particle size under 10nm to tune the fluorescence color and the peak wavelength [9], ternary sulfide based phosphors such as $Ba_2ZnS_3$:Mn (BZS), $SrGa_2S_4$:Eu (SGS), and $BaAl_2S_4$:Eu (BAS) have been focused on as the alternative phosphor nanoparticle materials which do not include harmful elements. Emission wavelength of these phosphor nanoparticles would be independent of the particle size since the doped impurities would be the origin of the fluorescence.

In this paper, these nanoparticle phosphors were coated by ZnO. The bandgap of ZnO is wider than CdSe and CdS, but smaller than ZnS and other phosphor materials. Coating of the

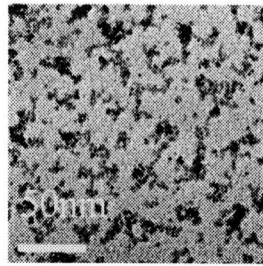

Figure 1  TEM image of BZS nanoparticles

nanoparticle is generally used to form the quantum confinement structure by utilizing the bandgap difference.  Because of the reversed bandgap difference, the quantum confinement was not expected for ZnS/ZnO, BZS/ZnO or other material systems.  The other function of the coating is to terminate the surface defect of the nanoparticle, which will result in an improvement of the optical properties.  Such effects were studied.  The nanoparticle was also sprayed on the glass substrate and thin layers were formed.  The optical properties of nanoparticle sprayed films were studied toward the application of Si solar cells.  Internal quantum efficiency of single crystal silicon solar cell is above 90% for most of the wavelength regions, but its value is around or less than 50 % in the UV to blue color region.  In order to solve the spectroscopic problem, red color transparent ink utilizing nanoparticle phosphors was applied to the surface of Si solar cells.  The red phosphor can convert the short wavelength light to the long wavelength light (which can be efficiently converted by the Si solar cell).  The role of the nanoparticle is to convert the UV portion of the incident light.  The concern for those wavelength converters (from UV to red) would be the internal loss of the long wavelength light; the long wavelength light might be absorbed in the red phosphor layer formed on top of the single crystal Si layer and not reach the solar cell surface.  The advantage of the nanoparticle phosphor is that nanoparticles would not efficiently absorb lights in the long wavelength region; hence nanoparticle phosphors may be a suitable material system to improve the spectroscopic response of single crystal silicon solar cells.

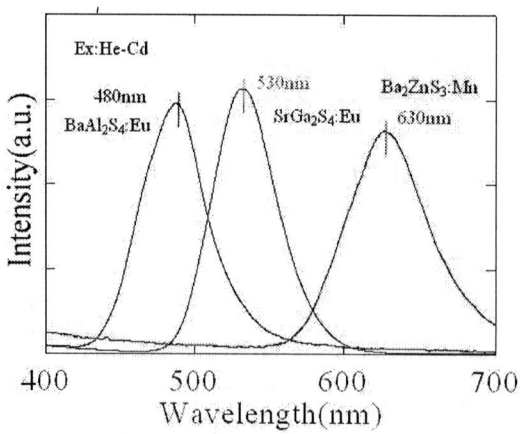

Figure 2  PL spectra of various nanoparticle phosphors

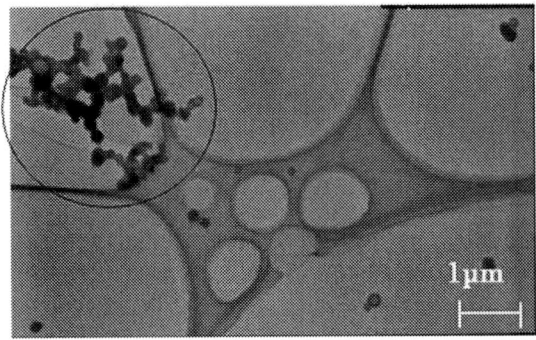

Figure 3 TEM image of ZnO nanoparticles without the core.

## EXPERIMENT

Starting powders of various phosphor materials were comminuted down to the nanometer size by the ball milling method. The particle size of the starting powders was about 1 micron. Some materials were grinded to the size of about 1 micron before loading into the milling machine. The milling was performed in a planetary micro milling machine (Fritsch P-7) using a stainless-steel pot containing the starting powder and $ZrO_2$ balls. The diameter of the milling ball was varied from 20 microns to 1 mm. The milling was performed for 10 to 60 hours with various rotation speeds. Those parameters were adjusted for each material. Prepared

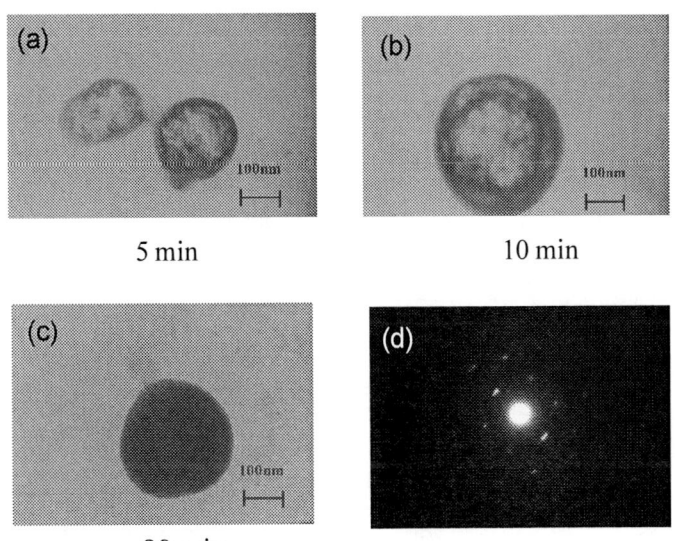

Figure 4 TEM image of the ZnO nanoparticle with various reaction times.
Diffraction pattern of the particle is shown in (d)

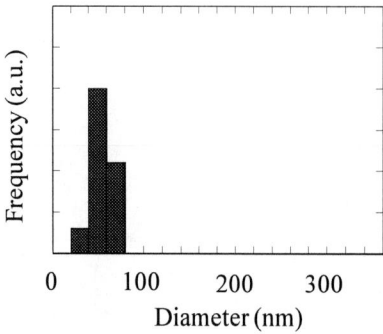

Figure 5 Particle size distribution of the ZnO nanoparticle.

nanoparticles and a zinc compound were mixed in an alcohol to form the coating layer and establish core/shell structures. The size distribution and the mole fraction ratio of the milled particle were evaluated using transmission electron microscopy (TEM: JOEL JEM2100F) and TEM- energy-dispersive X-ray spectroscopy (EDX) observations. The photoluminescence (PL) properties of nanoparticles were evaluated by irradiating the nanoparticles dispersed in solvents using a He-Cd laser (Ex: 325nm) at room temperature. Comminuted nanoparticles were dispersed in various kinds of alcohol, and the nanoparticle-containing alcohol was sprayed on the glass substrate using a spin coater. The particle distribution on the glass was studied using an atomic force microscopy (AFM), and its transmittance was measured in the wavelength region of between 400 nm and 800 nm. These films were then used as a light conversion layer for the single crystal Si solar cell.

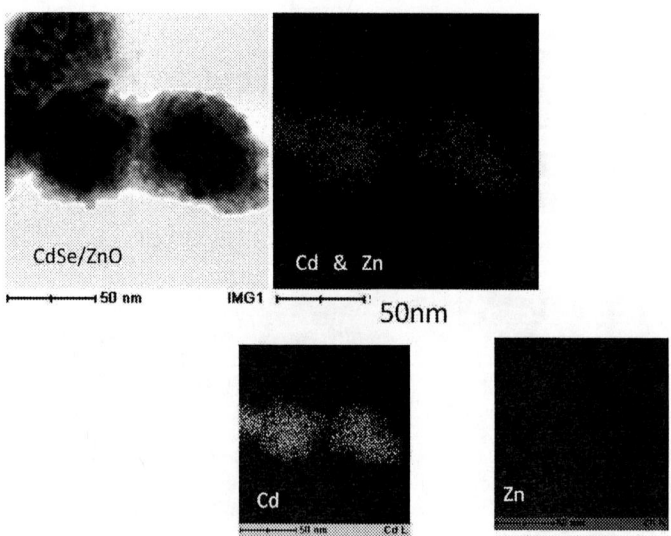

Figure 6 Real image and atom distribution profiles for ZnO coated CdSe nanoparticles.

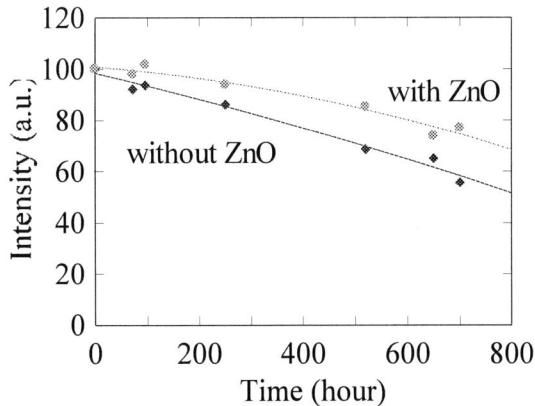

Figure 7 Luminescence intensity degradation of BZS and BZS/ZnO nanoparticles.

## DISCUSSION

Figure 1 shows a typical TEM image of milled BZS nanoparticles. As shown in the figure, particles of around 10 nm were obtained. These particles were well dispersed and it was clear that aggregation had been avoided by using a certain milling procedure. TEM-energy dispersive x-ray (TEM-EDX) was used to evaluate the stoichiometry of nanoparticle materials. Because of the particle size and detection limit of the TEM-EDX, precise analysis was difficult, but reasonable agreement of the stoichiometry between the starting material and the milled particles was achieved. Even after the mechanical milling, photoluminescence spectrum shapes were unchanged and emitted red light peaking at around 640 nm. The spectrum itself was not affected by the particle density, and the dominant peak remained at around 640 nm. The PL intensity was affected by the density of the sprayed nanoparticle; stronger PL signal was observed from the film that had the higher density of nanoparticles. Figure 2 shows the PL spectra of various nanoparticle phosphors. Red, green, and blue luminescences were observed from, BZS, SGS, and BAS, respectively, and the spectra were essentially the same as those of

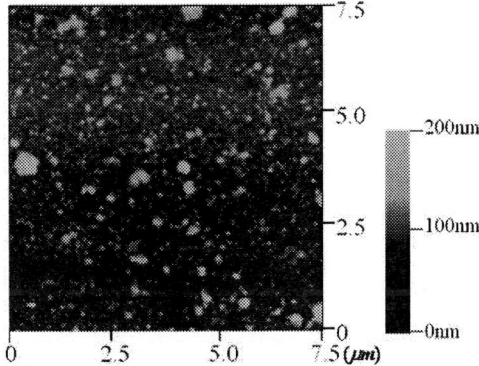

Figure 8 AFM image of a BZS film. (Sprayed particle density was about $4\times10^6$ particles/mm$^2$).

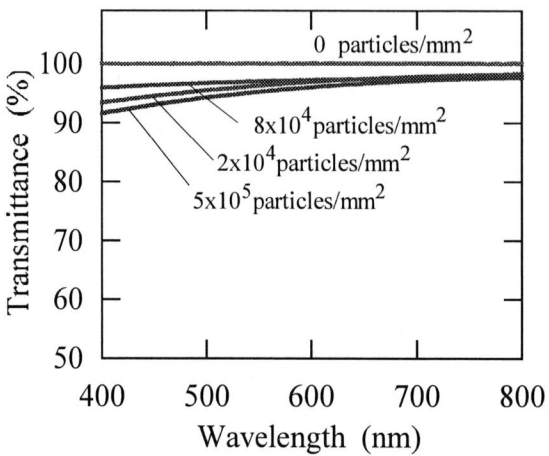

Figure 9 Transmittance spectra of BZS films.

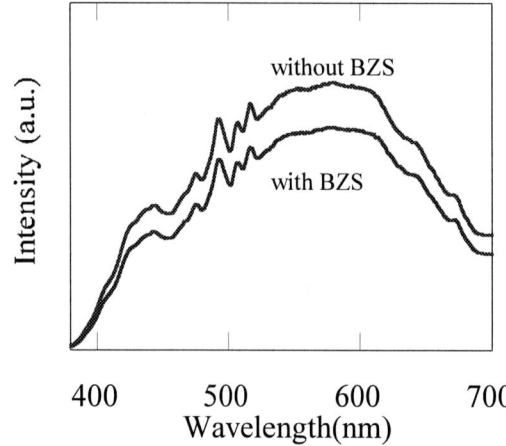

Figure 10  Spectra responses of a Si photodiode with and without BZS

the starting materials.  These results indicated that the optical quality of the starting material hasn't been seriously damaged by the ball milling.

Figure 3 shows the TEM image of the ZnO nanoparticles without the core material.  As can be seen from the figure, well dispersed nanoparticles could be prepared.  The TEM image was studied as a function of the reaction time.  Figure 4 shows the TEM image of the ZnO nanoparticle with various reaction times.  Solid nanoparticle formation was confirmed for the particle with a reaction time of 30 min, but weak contrasts at the center of particles were observed when the reaction time was about or less than 10 min.  The difference of contrasts was associated with the difference of the formed material.  The surface regions consisted only of ZnO when the reaction time was short, but the extended reaction time resulted in the formation of the solid ZnO nanoparticle.  The diffraction pattern of the solid ZnO nanoparticle is shown in Fig4(d).  As can be seen from the diffraction pattern, the particle was single-crystalline.  By changing the volume of the precursors, the nanoparticle size could be varied.  Figure 5 shows a typical particle size distribution of the formed ZnO nanoparticle.  The average size of ZnO was about 40 nm which would imply that the average thickness of ZnO core for the core/shell nanoparticle would be about 20 nm. The milled nanoparticle was coated by ZnO using this technique and TEM-EDX was used to characterize the coating profile. Figure 6 shows real image and atom distribution profiles for ZnO-coated CdSe nanoparticles. As can be seen from the figure, Cd and Se are distributed only inside Zn, giving the direct evidence for the formation of the ZnO coating layer around the CdSe nanoparticle.

Compared to the PL spectrum of the uncoated nanoparticles, the coated nanoparticles exhibited a slight peak shift and broadening, as well as a slight decrease of the peak intensity. The ZnO coating has affected the local strain field which has probably resulted in the PL peak shift. Since the bandgap of ZnO is smaller than those of phosphor materials used in this study, the light absorption in the coating layer has probably caused the decrease of the PL peak intensity. The essential characteristics of the nanoparticle, however, were not affected by the coating. PL intensity decay was measured as a function of time. Figure 7 shows the luminescence intensity degradation of BZS and BZS/ZnO nanoparticles as a function of the time. Statistical change in the fluorescence intensity of the nanoparticles immersed in a solvent was observed in the range of several hundred hours at room temperature. As shown in Fig.7, the fluorescence intensity decreased monotonically with the time, but ZnO coated BZS nanoparticles kept brighter fluorescence for a longer time than uncoated BZS. The fluorescence intensity of BZS /ZnO nanoparticles 700 hours after the coating treatment showed the fluorescence intensity was 1.4 times stronger than that of uncoated BZS nanoparticles. SGS and BAS showed similar results. The capping of the unstable sulfide nanoparticles with chemically stable ZnO has resulted in the improvement of the chemical stability of the entire nanoparticle.

Figure 8 shows the AFM image of the nanoparticle sprayed on the photodiode surface. The nominal density of the particle sprayed on the surface is about $4 \times 10^6$/mm$^2$. It was revealed that well dispersed nanoparticles were uniformly covering the surface. Judging from the surface image, substantial areas were not covered by nanoparticles, which contributed to the high transparency of the thin film. There are many aggregated particles whose size is on the order of submicrons. These aggregated particles were probably formed during the drying process. The elimination of the aggregation would be critical to not damage the transmittance of the film and the resulting sensitivity of the photodiode. Transparency of the nanoparticle film is shown in Fig. 9 as a function of the wavelength (400 nm to 800 nm). The degree of transparency was affected by the amount of sprayed nanoparticles, but the transmittance all over the visible region was above 90%. When the particle density was about $8 \times 10^4$ /mm$^2$, transmittance around 400 nm to 440 nm was slightly lower than other wavelength regions, and its value is around 95 %. On the other hand, the transmittance change became noticeable in the short wavelength region once the density of the particle has increased; the transmittance was about 90 % at 400 nm for the film whose particle density was about $4 \times 10^5$ /mm$^2$. The amount of aggregated nanoparticles was responsible for the degradation of the transmittance.

Figure 10 compares the spectroscopic response of the photodiode with and without the sprayed nanoparticle layer. The measurement was performed at room temperature. The white light was shone onto from the surface of the photodiode. As can be seen from the figure, the spectroscopic response of the sample with the nanoparticle layer was similar to that of the sample without the nanoparticle layer. However, spectroscopic response without the nanoparticle layer exhibited higher sensitivity in most of the wavelength region. At the region around 400 to 440 nm, the photo response was almost the same for both samples. Since the spectroscopic response was not lowered in the wavelength region where the photo-absorption by the BZS was expected, the wavelength conversion was effectively performed for this region. The lowering of the spectroscopic response at the other wavelength region was probably associated with the aggregated nanoparticles on the photodiode surface which resulted in the decrease of transmittance in longer wavelength regions. From the above results, it is clear that nanoparticle phosphor films could be used to improve the conversion efficiency of the single crystal Si solar cells.

## CONCLUSIONS

CdSe, ZnS, $Ba_2ZnS_3$:Mn, $SrGa_2S_4$:Eu, and $BaAl_2S_4$:Eu nanoparticles were prepared by the ball-milling method. Several-nanometer-size stoichiometric and dispersed nanoparticles were achieved. Bright luminescence was observed from most of the nanoparticle phosphors. ZnO-coating was performed and a uniform coating layer was formed on the phosphor nanoparticles. The ZnO-coated core/shell nanoparticles exhibited improved chemical stability. BZS nanoparticles sprayed on top of a Si photodiode exhibited that wavelength conversion occurs.

## ACKNOWLEDGMENTS

This work is supported in part by Waseda University High Tech Research Center Project, Organization for University Research Initiatives, and MEXT KIBANKEISEI.

## REFERENCES

1. B. O. Dabbousi, J. Rodriguez-Viejo, F. V. Mikulec, J. R. Heine, H. Mattoussi, R. Ober, K. F. Jensen, and M. G. Bawendi, J. Phys. Chem. B **101**, 9463-9475 (1997)
2. 3. M. A. Hines and P. Guyot-Sionnest, J. Phys. Chem. **100**, 468-471 (1996)
3. 4. H. Borchert, D. V. Talapin, C. McGinley, S. Adam, A. Lobo A. R. B. de Castro, T. Moller, and H. Weller, J. Chem. Phys. **119.3**, 1800-1807 (2003)
4. X. Peng, M. C. Schlamp, A. V. Kadavanich, and A. P. Alivisatos, J. Am. Chem. Soc. **119**, 7019-7029 (1997)
5. I. Mekis, D. V. Talapin, A. Kornowski, M. Haase, and H. Weller, J. Phys. Chem. B **107**, 7454-7462 (2003)
6. R. N. Bhargava, D.Gallagher, X.Hong, and A.Nurmikko, Phys. Rev. Lett. **72**, 416 (1994)
7. S. Ishizaki, Y. Kusakari, and M. Kobayashi, Mater. Res. Soc. Symp. Proc. **829**, B2221-2225 (2005)
8. S. Hamaguchi, S. Ishizaki, and M. Kobayashi, Korean Phys. Soc. 53,5 (2008) 3029-3032
9. S. J. Rosenthal, J. McBride, S. J. Pennycook, and L. C. Feldman, Surface Sci. Rep. **62**, 111-157 (2007)

Mater. Res. Soc. Symp. Proc. Vol. 1394 © 2012 Materials Research Society
DOI: 10.1557/opl.2012.696

### Diffusion of ion implanted indium and silver in ZnO crystals

Faisal Yaqoob[1] and Mengbing Huang[2,*]

[1] Department of Physics, State University of New York, 1400 Washington Ave, Albany, NY, 12222

[2] College of Nanoscale Science and Engineering, University at Albany, State University of New York, 1400 Washington Ave, Albany, NY, 12222

* Email: mhuang@albany.edu

# ABSTRACT

We report on diffusion behavior for ion implanted indium and silver atoms in ZnO crystals. Both In and Ag ions were implanted at room temperature at 7-10° relative to c-axis to avoid channeling effects during implantation. In ions were implanted at four different energies (40, 100, 200, and 350 keV, respectively) and doses ($4.20 \times 10^{13}$, $6.70 \times 10^{13}$, $8.10 \times 10^{13}$ and $3.10 \times 10^{14}$ /cm$^2$, respectively), resulting in a total dose of $5 \times 10^{14}$ /cm$^2$. For another set of ZnO samples, Ag ions were implanted at energies 30, 75, 150, and 350 keV at doses $3.3 \times 10^{13}$, $4.2 \times 10^{13}$, $8.3 \times 10^{13}$ and $3.4 \times 10^{14}$ /cm$^2$, respectively, to reach a total dose of $5 \times 10^{14}$ /cm$^2$. Both In and Ag implants resulted in a uniform concentration profile of the implanted dopants from surface to depth ~ 150 nm. The samples were annealed for 30 minutes at temperatures between 850-1050 °C in an oxygen gas flow. The distributions of In and Ag atoms, either aligned or nonaligned along the crystalline directions, were measured by Rutherford backscattering combined with ion channeling. The diffusivities for nonaligned (interstitial) and aligned (substitutional) dopants atoms were determined to vary with annealing temperature via the Arrhenius relationship. The diffusion activation energies ($E_a$) along the <10-11> direction for substitutional impurity atoms were lower than those for interstitial dopants atoms e.g., in the case of In, $E_a$ ~ 1.52 eV for <10-11> aligned In atoms and $E_a$ ~ 2.61 eV for interstitial In atoms between <10-11> atomic rows and in the case of Ag, $E_a$ ~ 1.77 eV for the interstitial Ag atoms between the <10-11> atomic rows and 1.11 eV for <10-11> aligned Ag atoms. The diffusion activation energies showed a different trend for the two dopants as measured along the <0001> crystalline direction. For Ag implanted in ZnO, the activation energy of $E_a$ ~ 0.91 eV for the aligned Ag atoms along <0001> direction and $E_a$ ~ 1.55 eV were found for the interstitial Ag atoms, whereas in the case of In along the <0001> direction, the interstitial In was found to migrate with a higher activation energy ($E_a$ ~ 1.78 eV) than the substitutional In ($E_a$ ~1.42 eV). These results will be compared with first-principle calculations for understanding the energetics of defect formation and migration in both n- and p-type doping cases.

# INTRODUCTION

ZnO is one of the most studied materials in the last decade or so. It has wide applications in ultraviolet light emitters and detectors, high-power and high-frequency electronic devices, sensors, and transparent conductors [1]. Its high exciton binding energy (~ 60meV, compare to ~ 25meV for GaN), its high resistance to radiations, high reactivity to wet etching, smaller dislocation density and availability in bulk single crystal form makes it strong competitor to GaN and other wide bandgap semiconductors [2-4]. Being a wide-band-gap material (3.4eV), it becomes a good candidate for UV optoelectronics [5]. Its transparency to visible light provides opportunities to develop transparent electronics, UV optoelectronics, and integrated sensors all from the same material system cheaper and less poisonous than indium tin oxides. To achieve many of these applications, ZnO still has to overcome the issue of reproducible p-type doping. ZnO is naturally an n-type semiconductor because of the low formation energies native defects like oxygen vacancies and zinc interstitials. Like all other wide band gap semiconductors, (ZnSe, CdS, GaN, etc), it is very difficult to obtain p type doping in ZnO. A number of dopants have been studied, but so far reproducible p type ZnO has been not reported [6, 8, 10, and 11]. Recently Ag is studied as a potential p type candidate [12-15]. First principle calculations have shown very low formation energies for Ag at zinc substitutional sites compared to interstitial sites [16].

In this study, we investigate the diffusion activation energies of indium and Ag in ion implanted ZnO crystals. We will study the formation energies of n-and p-type dopants in ZnO crystals.

# EXPERIMENTAL DETAILS

Single ZnO crystals, e and f, each of size of $1 \times 1$ cm$^2$, were used for this experiment. Both these samples were ion implanted with silver (sample e) and indium (sample f) to a total dose of $5 \times 10^{14}$ /cm$^2$. Each of the implantation was a multiple energy implantation at four different energies 30, 75, 150, and 350 keV, and doses $3.3 \times 10^{13}$, $4.2 \times 10^{13}$, $8.3 \times 10^{13}$ and $3.4 \times 10^{14}$ /cm$^2$, respectively for Ag and at four different energies 40, 100, 200, and 350 keV, and doses $4.20 \times 10^{13}$, $6.70 \times 10^{13}$, $8.10 \times 10^{13}$ and $3.10 \times 10^{14}$ /cm$^2$, respectively for In, resulting in a uniform concentration profile of Ag and In from the surface to depth $\sim 150$ nm. The TRIM simulations for silver depth profile and the implantation energies are shown in Fig. 1. A very similar depth profile was done for indium in ZnO. During implantation, the samples were tilted by $\sim 7°-10°$ relative to the perpendicular ion beam to minimize channeling effects during implantation. These samples were annealed at temperatures 850-1050 °C for 30 minutes in an oxygen gas flow.

The distribution of dopants (Ag and In atoms), either aligned or nonaligned along the crystalline directions, were measured by Rutherford backscattering (RBS) combined with ion channeling. Helium beam with energy 3.05MeV, surface charge dose of 8μC and beam intensity $\sim 0.50$nA – 1.00nA for random and aligned respectively, was used for RBS/Channeling experiment The samples were analyzed along two crystallographic directions, <0001> and <10-11>. When the projectile He beam is steered between the <0001> atomic rows, the helium beam is reflected only from the surface atoms or atoms between the atomic rows. In which case it can miss the Ag atoms in the shadow cone behind a surface atom, and the in-channel fraction of Ag wouldn't reflect the total or most of the Ag concentration within the crystal [19]. Therefore channeling along <10-11> crystallographic direction is also done so that the He beam can be backscattered from most of the Ag atoms either within the atomic rows on interstitial positions or on substitutional lattice positions. The samples were analyzed along both <10-11> and random directions. Although pure random direction can only be found for amorphous material, we continuously rotated the sample to get an accurate random spectrum.

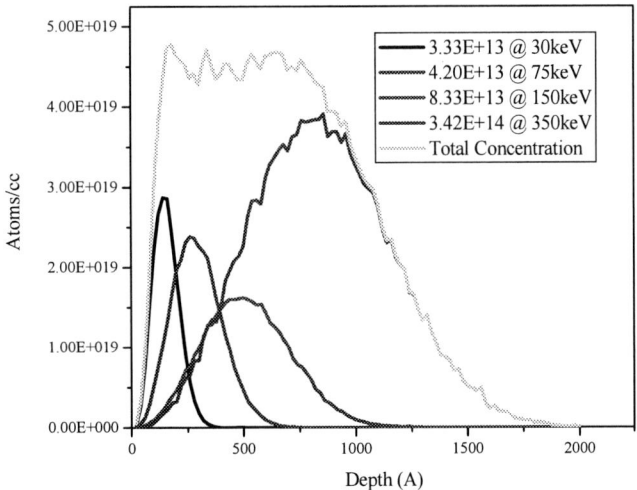

Fig. 1: TRIM simulation for Ag implantation in ZnO crystals.


**Mathematical Model**

The implanted Ag and In is diffused out from the implanted region as the samples are annealed. This can be seen by comparing the RBS spectrum for the as implanted sample with one of the annealed samples as shown in the insets of Fig 2. To model this diffusion we use the standard Fick's law of diffusion with an additional term for generation and dissociation of defect complexes. Let *[A]* be the concentration of dopants and *[B]* be the vacancies or point defects available to them, then the time rate of change of the dopant concentration can be modeled as:

$$\frac{\partial [A]}{\partial t} = \frac{\partial}{\partial x}\left( D_A \frac{\partial [A]}{\partial x} \right) - \frac{\partial [AB]}{\partial t},$$

where, the last term on the right side is the generation and dissociation term given by:

$$\frac{\partial [AB]}{\partial t} = K[A][B] - \nu [AB]$$

$D_A$ is the diffusivity constant, $K$ is the generation constant and $\nu$ is the dissociation frequency. The generation constant $K$ is related to the capture radius $R$ and the diffusivity constant $D_A$ by $K = 4\pi R D_A$. If $B_{tot}$ is the number of traps, then the concentration of traps left after the formation of complexes between the defects and the dopants is: *[B] = B_{tot} –[AB]*. We simplified the above equations by using the fact; that the ratio of the integrated dose of any of the annealed sample to the integrated dose of the asimplanted sample is small number. And then we worked out the diffusivity constant satisfying the Arrhenius equation, $D = D_0\,exp(-E_a/KT)$, where $E_a$ is the migration activation energy.

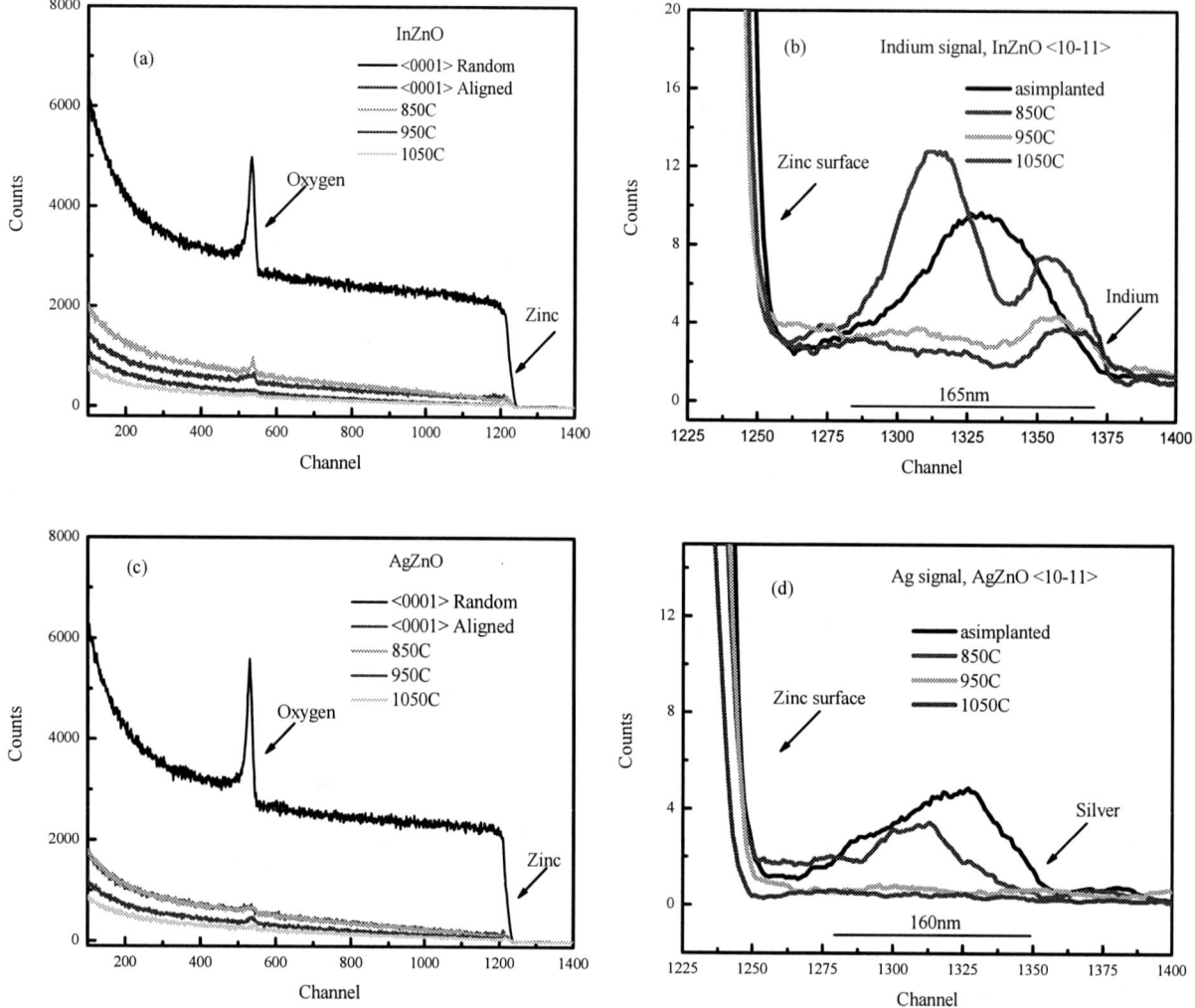

Fig. 2: RBS spectra for (a): indium and (c): silver implanted ZnO crystals measured along <0001> direction for aligned and random cases, (b): indium signal measured along the <10-11> direction for the asimplanted and annealed samples, and (d): silver signal measured along the <10-11> direction for the asimplanted and annealed samples.

## RESULTS AND DISCUSSIONS

The Rutherford backscattering and ion channeling spectra for indium and silver implanted ZnO crystals are presented in Fig. 2, (a) and (b) respectively. The insets in (a) and (b) are showing the indium and silver signal as measured along <10-11> directions. Both indium and silver were implanted to a similar depth, same dose and annealed under similar conditions and in both cases the crystal lattice recovers. The damage to the oxygen lattice is much higher in each case as explained in ref [9]. The Zn and Ag lattices recover to a same fraction in each case at the highest annealing temperature, but oxygen lattice damage is significantly larger in indium implanted crystals as compared to silver implanted samples when both are annealed at 950 °C. This could be due the high affinity of indium for oxygen than for zinc. Indium forms complexes with oxygen resulting in higher RBS yield for oxygen at 950 °C. As the samples are annealed, a very sharp RBS peak is formed in indium implanted crystals in the near surface region. This is due the diffusion of indium atoms in to the <0001> atomic rows. This near surface peak is reduced as the samples are annealed to higher temperatures; also the loss in RBS yield of indium in below surface region indicates that the indium is diffusion of the implanted depth to the bulk and to the near surface region. Interestingly, the surface peak does not disappear even at the highest annealing temperature of 1050 °C, showing that the bonds between the indium oxygen complexes are holding together, though some of the bonds are broken and the RBS yield has decreased compared to 850 °C annealed sample.

The thermal process also causes diffusion of the substitutional and interstitial impurity atoms in the silver implanted crystals but there is no surface peak formation, see Fig. 2 (d). The RBS yield is significantly decreased at 950 °C and 1050 °C showing the diffusion of Ag atoms, for example there were 30.2% of the Ag atoms on Zinc substitutional ($S_{zn}$) in the asimplanted case which increased to 61.6% after annealing at 1050 °C. In the indium implantation case there was 70.6% of the indium atoms on substitutional sites in the asimplanted sample which becomes 88.5% after annealing at 1050C.

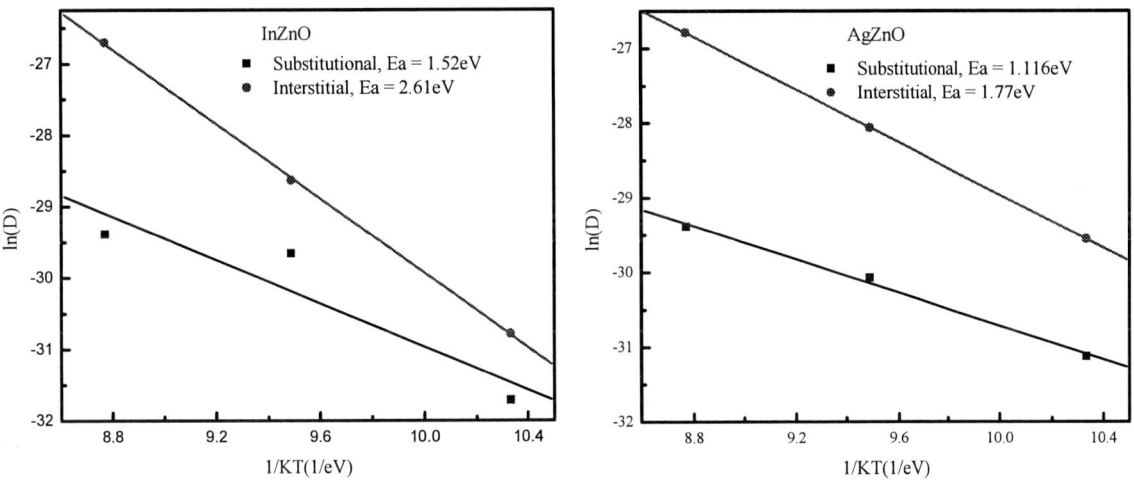

Fig. 3: Effective migration energies for interstitial and substitutional indium and silver dopants in ZnO crystals.

Fig. 3, shows the *1/KT* vs the log of diffusivity constant *D* in an Arrhenius manner. The effective activation energies are obtained by making liner fits to the data. In vacancy assisted mechanism, the substitutional dopants diffuse by exchange between the Zn vacancy and the dopants. In this process energy is required to create a vacancy and to move an impurity atom between these vacant sites. Since during ion implantation large numbers of vacancies are created at the surface and interstitials are created at the implantation depth, so the effective activation energy measured in this work is possibly related to the migration energy of impurity atoms in ZnO crystals. The difference between the substitutional and interstitial dopants migration energies can be explained by the fact that to knock out an atom from a substitutional site and diffuse it through the crystal requires large amount of energy than the energy required to knock out an interstitial atom and its diffusion through the crystal. The substitutional indium and silver atoms are found to be stable than the interstitial indium and silver atoms, for example the value of migration energy for the substitutional indium is 1.52eV compared to 2.61eV for the interstitial indium atoms. Similarly the migration energy of the substitutional silver is 1.11eV, whereas the interstitial silver atom moves with 1.77eV energy, i.e., the interstitial impurity atoms are fast diffuser in each case. The substitutional indium atom has been found to migrate within the energy range of 0.63 - 2.7eV, both theoretically and experimentally by various groups [18-21]. The migration energies for substitutional silver atoms are found to be lower than those found in the literature (1.43 – 2.06eV) where are the interstitial silver in our case is migrating with higher activation energy compared to the first principle calculation value of 0.98eV [17, 22]. The substitutional Ag is difficult to diffuse compare to substitutional In. This is possibly due to the complex formation between indium and oxygen. An indium atom substituting Zn at the tetrahedral position and bonded to oxygen to form $In_2O_3$ or InO type complexes may be easily diffused in comparison to Ag atom substituting the Zn site. More detailed study and experiments are required to understand this phenomenon on atomic level.

# CONCLUSIONS

In conclusions, we found that despite having greater activation energies, the interstitial impurity atoms are found to diffuse much faster than the substitutional impurity atoms. Interstitial In and Ag atoms exhibit higher activation energies (2.61eV vs 1.77eV) and diffusivity in ZnO. The substitutional Ag atoms are found to be more stable than the substitutional In atoms. Further work is under process to understand this phenomenon on atomic scale.

**REFERENCE**
1. D. C. Look, Material Science and Engineering B80, 383, (2001).
2. Ü. Özgür, *et al* J. Appl. Phys. 98, 041301 (2005).
3. Anderson Janotti and Chris G Van de Walle, Rep. Prog. Phys. 72 (2009), 126501.
4. D.C.Look, Appl. Phys. Lett. 75, 811 (1999).
5. S. J. Pearton, et al., Superlattices and Microstructures 34, 3-32, (2003).
6. T. Minami, et al. Jpn. J. Appl. Phys., Part 2 **24**, L781 (1985).
7. W. Walukiewicz, Phys. Rev. B **50**, 5221 (1994).
8. C. G. Van de Walle,et al. Phys. Rev. B **47**, 9425 (1993).
9. Effects of Hydrogen Ion Implantation on Structural Properties of Silver Implantation in ZnO Crystals, Faisal Yaqoob, Mengbing Huang, Submitted to MRS Fall 2011, Proceedings.
10. S. B. Zhang, et al, Phys. Rev. B, 63, 075205, (2001).
11. D. C. Look, et al, Phys. Rev. Lett. **82**, 2552 (1999).
12. Hong Seong et al., Applied Physics Letters, 88, 202108, (2006).
13. E. Rita et al., Hyperfine Interactions, 158, 395, (2004).
14. Jiwei Fan and Robert Freer, J. Appl. Phys. 9, 4795, (1995).
15. Isao Sakaguchi, et al., Journal of Ceramic Society of Japan, 118[3], 217, (2010).
16. Yanfa Yan, et al., Applied Physics Letters, 89, 181912, (2006).

17. G. Y. Huang et al., J. Phys.; Condens. Matter 21, 345802, (2009).
18. G. Y. Huang et al., J. Phys. 105, 073504, (2009).
19. T. Nakagawa et al., Jpn. J. Appl. Phys. 47, 7848, (2008).
20. D. G. Thomas, J. Phys. Chem. Solids, 9, 31, (1959).
21. Q. Wan et al., Optical Materials, 30, 817, (2008).
22. I. Sakaguchi, et al, Nucl. Instr. and Meth. in Phys. B 206, 153, (2003).

**Mater. Res. Soc. Symp. Proc. Vol. 1394 © 2012 Materials Research Society**
**DOI: 10.1557/opl.2012.802**

## Effect of Low Power Deposition and Low Oxidation Temperature on the Interfacial and Structural Properties of sputtered HfO₂ Gate Dielectrics

Auxence Minko[1] Gustavo S. Belo[1] Sergei Rudenja[2] and Douglas A. Buchanan[1]

[1]Electrical & Computer Engineering Department, University of Manitoba, 75 Chancellors circle, Winnipeg, MB R2H0C6, Canada.
[2]Department of Chemistry, 144 Dysart Rd, Parker Building, University of Manitoba, Winnipeg, MB R3T 2N2 Canada.

### ABSTRACT

Hafnium dioxide gate dielectrics, prepared by DC magnetron with low-power sputtering deposition followed by a low-temperature thermal oxidation, show greatly improved interfacial and electrical properties. Ellipsometry and X-ray photoelectron spectroscopy (XPS) measurements show a good stoichiometric $HfO_2$ thin films with a refractive index of 1.9 and an Hf:O ratio of 1:2. The results obtained after analysis, quantification and calculation through XPS depth profile method, angle resolved XPS and interface modeling by XPS data processing software suggest a development of a complex three layer dielectric stack, including hafnium dioxide layer, a narrow interface of hafnium silicate and broad region of oxygen diffusion into silicon wafer. The measured dielectric constant of the $HfO_2$ was about 22. The film band-gap was found to be ~ 5.2 eV.

### INTRODUCTION

The continuous scaling of the $SiO_2$ layer for advanced CMOS technologies has reached its physical limit which has led to excessive gate leakage and reliability issues. To address this problem, high-κ materials have been investigated as potential $SiO_2$ substitutes. Hafnium dioxide ($HfO_2$) is being considered as one the most promising candidate, due to its superior physical and electrical properties. These include, naming a few, a high permittivity, band offsets with Si greater than 1, and good thermal stability [1]. However, the deposition of $HfO_2$ using atomic layer deposition (ALD), chemical vapor deposition (CVD) and sputtering techniques can lead to the formation of an interfacial layer between $HfO_2$ and silicon. This interlayer creates an additional contribution to effective oxide thickness (EOT) and negates the effect of using a high-κ dielectric. Many and various efforts have so far been made to overcome this degradation.

In this work, a method to deposit a high quality film of $HfO_2$, while maintaining a narrow interface layer thickness, is investigated. The characterization of the resulting gate dielectric (hafnium dioxide and interface layer) and their properties have been investigated using high resolution TEM, X-ray photoelectron spectroscopy, ellipsometry and capacitance-voltage measurement.

The TEM images reveal a small interface layer while the XPS depth profile data, analyzed through XPS MultiQuant program, show that the gate dielectric is actually composed of a three layer structure that includes a stoichiometric $HfO_2$ layer, a thin layer of $HfSiO_4$ and a thin layer of $SiO_2$ at the silicon interface. Optical properties such as refractive and absorption indexes lead to a suitable bandgap of 5.2 eV and the capacitance voltage measurement shows a high relative permittivity.

**EXPERIMENT**

Silicon (100) p-type wafers with a resistivity 1-10 Ω·cm were used in this study. The conventional (RCA) technique was performed in order to pre-clean the silicon surface. This step was followed by the removal of the native oxide in a buffered HF (1%) solution for 3 minutes. The fabrication of $HfO_2$ thin films was done using two different approaches. For the first sample (S1), the $HfO_2$ thin film was deposited by the DC magnetron sputtering with a power density of 2.46 W/cm$^2$ from a high-purity Hafnium (Hf) target in a mixture of high-purity oxygen (O) and argon (Ar) at room temperature. A thin layer of Hafnium metal was sputtered onto the two separate silicon wafers, S2 and S3, at a power density of 1.23 and 2.46 W/cm$^2$, respectively. For both samples S2 and S3, Argon was used as sputtering gas. The pressure during the deposition was kept constant at 12 mTorr. After the Hf-deposition, the separate pieces of the thin metal films (S2, S3) were thermally oxidized at 400°C, 500°C and 600°C for 1 hour in $O_2$ (99.999).

Both deposition power and the post-deposition oxidation annealing temperature were used to investigate their effects on the formation of the interfacial layer. Separate pieces of the sample S2 were annealed for an hour in oxygen at 400°C, 500°C and 600°C. Using the temperature that lead to the best result, sample S3 was annealed for comparison of the effects of the deposition power on the interfacial layer formation.

X-ray photoelectron (XPS) measurements were performed at a base pressure of $1\times10^{-9}$ Torr using a Kratos Axis Ultra spectrometer with the monochromatic Al $K_\alpha$ (hv = 1486.6 eV) source. A survey spectrum was first taken with a pass energy of 160 eV. All high resolution spectra were collected with a pass energy of 40 eV. A depth profile analysis was performed using Ar$^+$ sputtering at 500 eV in a floating voltage mode, over a rhombic area, 2 mm on a side. Gate dielectric layers quantification was performed using the XPS MultiQuant program [2]. A spectroscopic ellipsometer ESM 300J (A. Woollan Co., Inc.) was used to characterize the thickness, the refractive index and the band-gap of the film. Cross-sectional images of different stacks were taken using the Transmission Electron Microscope (JEOL 2010F TEM). High frequency capacitance–voltage was measured using a HP 4284A Precision LCR.

**DISCUSSION**

In Figure 1, high resolution TEM (HRTEM) images of $HfO_2$/Hf-Si-O/Si gate stacks resulting oxidation at a) 400 °C, b) 500 °C and c) 600 °C of sample S2, are shown. These TEM images show the impact of the oxidation temperature on the interlayer thickness. It is clearly seen that the interlayer thickness ($SiO_x$ in Figure 1) increases with the increasing oxidation temperature. It was found that the interfacial layer shrinks from ~3.5 nm at 600°C to ~1.1 nm at 400°C. Given that the Hf-O bond has a higher bond enthalpy (801.7 kJ/mol) than Si-O (799.6 kJ/mol), the hafnium should have a higher affinity to oxygen than silicon [3]. However, the diffusivity of atomic oxygen in silicon at 400°C ranges from ~$2\times10^{-20}$ cm$^2$/s to ~$1\times10^{-19}$ cm$^2$/s [4], which is equivalent to a diffusion length ranging from ~1.7 Å to ~3.8 Å for 60 min. The diffusion length is given by equation

$$L = 2\times\sqrt{(D\times t)} \qquad (1)$$

Consequently, a lower oxidation temperature can be employed to oxidize the Hf metal layer while limiting the diffusion of atomic oxygen into the silicon.

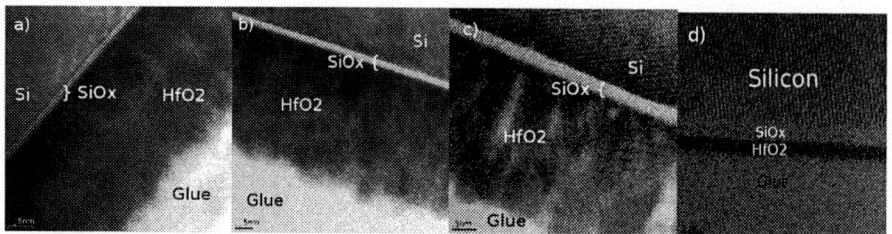

**Figure 1.** Cross-sectional high-resolution TEM images for $HfO_2/SiO_x/Si$ gate stack, after oxidation at a) 400°C, b) 500°C and c) 600°C. d) $HfO_2$ film deposited in a mix of $O_2$ and Ar gases at 2.46 W/cm$^2$.

In contrast, to the thin interfacial layer found for S2, the sample (S1) that was sputtered from an $HfO_2$ target in O2 shows a thick interfacial layer as shown in the Figure 2(d). This process led to the formation of $HfO_2$ but also resulted in the growth of ~5 nm thick, interfacial layer without any post-deposition annealing. Sample S3, where metallic hafnium was sputtered in Ar at the same power as sample S1, was annealed in oxygen at the lowest temperature (400°C) producing a thicker interfacial layer than S2. From TEM data, the interlayer thickness was found to be similar to S1 interfacial layer (~5 nm) – i.e the film sputtered from the $HfO_2$ target. This suggests that the higher sputtering power (2.46 W/cm$^2$) used during the sputtering deposition on both S1 and S3, independent of the target, facilitated the diffusion of the metallic Hf into Si, creating a broad region which contained both Hf and Si atoms, just below the deposited Hf layer. Implanted Hf atoms could then have been favored the formation of the interfacial layer during the oxidation process.

The composition and the stoichiometry of the hafnium oxide film (S2), annealed at 400°C, was analyzed using data collected from XPS.

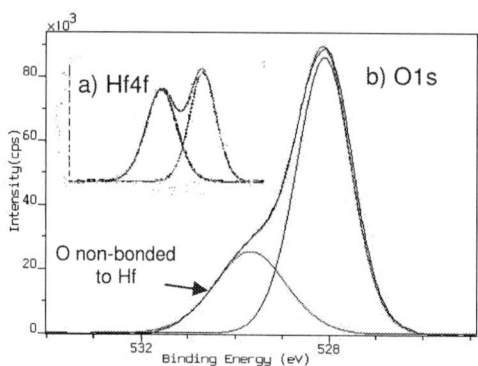

**Figure 2.** Core level spectra, a) Hf-4f and b) O-1s, for Hf sputtered film (S2) oxidized at 400°C.

In Figure 2, the core level spectra (Hf-4f and O-1s) of the Hf sputtered film (S2) oxidized at 400°C are shown. Two chemical states of Hf and two of O were detected in each of the core level spectra. From the XPS data, the different bonding configurations (i.e. oxygen bound to hafnium or another material) [5] can be differentiated. The O/Hf ratio can be accurately calculated from the spectra data, once the O not bounded to Hf is taken in account. From this data, the composition of the film was found to have a ratio of ~ 2, i.e. $HfO_2$.

The composition of the interlayer was investigated in a previous report [6]. In that study, the depth profiles of the oxide stack were shown (see Figure 3 below) in which the integrated intensity (adjusted by the relative sensitivity factors), was used to measure the composition profiles. In Figure 3, these intensities are shown as a function of the sputtering time in (sec). The profile was performed on a sample built, using the same deposition technique as S1. The sputtering removed the film at a rate of ~12.3 Å/min.

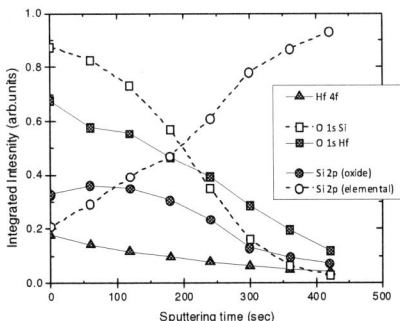

**Figure 3.** XPS depth profile of the oxide stack for samples deposited using the same recipes as samples S1.

In Figure 3, the gradual removal of the oxide stack and the transition from oxides to the silicon substrate are shown. The depth profiling confirms the extension of the $SiO_x$ region into the silicon wafer. Four chemical states were detected in the area below $HfO_2$ (Si 2p oxidized, Si 2p metallic and O 1s and Hf 4f ). These results suggest that the oxide stack is made of a complex three layers dielectric stack, $HfO_2$, Hf-Si-O, and $SiO_x$. The quantification of different layers was done using the MultiQuant software [6]. This XPS data processing allows successful quantification of not only homogeneous samples and their overlayers, but also of structured or layered films. It may also be used to analyze the composition of the surface by applying various equivalent structural models, for example, by varying the number of the layers at the interface and their stoichiometry. The data shown in Table I, suggests that a three-layered plane model consisting of a $HfO_2/HfSiO_4/SiO_2$ stack produced the most consistent result [6].

| Chemical composition of layer | Layer 1 | Layer 2 | Layer 3 | Bulk |
|---|---|---|---|---|
| | $HfO_2$ | $HfSiO_4$ | $SiO_x$ | Si |
| Molecular weight | 210.5 | 270.6 | 60.1 | 28.1 |
| Stoichiometry of layers | | | | |
| Si | | 1 | 1 | 1 |
| Hf | 1 | 1 | | |
| O | 2 | 4 | 2 | |

**Table I.** The molecular weight, stoichiometry and corresponding values used for the modeling of a three-layered oxide stack on a plane by MultiQuant software

Using spectroscopic ellipsometry, the $HfO_2$ film (S2) was found to have a thickness of ~23 nm and an average refractive index of ~1.9 in the visible range. Figure 4 shows the measured refractive index (n) and absorption index (k) of the film.

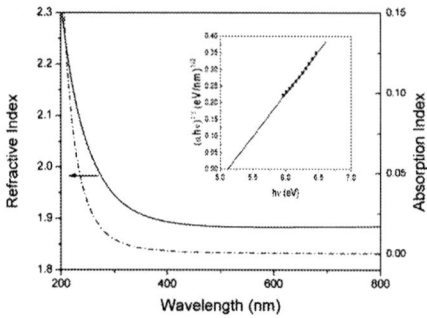

**Figure 4.** Refractive and absorption indexes of $HfO_2$ film from Hf sputtering and oxidized at 400°C for 60 min. The inset shows the extracted energy band-gap value of the material

The refractive index data fitted with the Cauchy diffraction model given by

$$n(\lambda) = A + B / \lambda^2 + C / \lambda^4 \qquad (2)$$

where $A = 1.89$, $B = 5.02 \times 10^{-3}$ nm$^2$ and $C = 8.93 \times 10^{-4}$ nm$^4$. The value of the bandgap was determined using:

$$(\alpha \upsilon) = B'(h\upsilon + Eg)^\eta \qquad (3)$$

where $\alpha = (4\pi k/\lambda)$ is the absorption coefficient of a photon at energy hv, B' is a constant and $\eta$ is dependent upon the optical transition; for amorphous dielectric films, $\eta = 2$ [7]. In the inset of Figure 4, the photon response, $(\alpha h v)^{1/2}$, is shown as a function of energy (hv) curve, from which the bandgap of the deposited $HfO_2$ was determined to be ~5.2 eV. This value is consistent with the values reported in the literature [7].

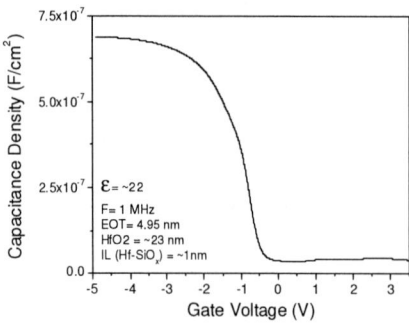

**Figure 5.** High frequency C-V of a 23 nm $HfO_2$ film (S2) oxidized at 400°C for 60 min.

Given the XPS and TEM results gave reasonable thin films, capacitors were fabricated (S2) by lithographically patterning of thermally evaporated aluminum. This was followed by a forming gas ($H_2$:$N_2$ 5:95) anneal (FGA) at 400°C for 30 min for interface state passivation. During the FGA, a further increase of the interface layer was unlikely because of the low oxygen content in the $H_2$:$N_2$. The high-frequency (1 MHz), capacitance-voltage (C-V) data of the $HfO_2$ capacitors (S2) are shown in Figure 5.

The total measured capacitance density of the stack, in accumulation, was found to be ~7×10⁻⁷ F/cm², corresponding to an equivalent oxide thickness (EOT) of ~5 nm. Using (4) below, and assuming that the interface layer is mostly $SiO_2$, the calculated relative permittivity of the stack $\varepsilon_{stack}$ is ~19 while $\varepsilon_r$ of $HfO_2$ is ~22.

$$1/C_{ox} = 1/C_{HfO_2} + 1/C_{SiO_2} \qquad (4)$$

These results demonstrate the ability of sputtering technique to fabricate a high quality $HfO_2$ film with a narrow interface layer when it is combined with the appropriate post-deposition annealing conditions.

## CONCLUSIONS

High quality $HfO_2$ thin films have been successfully deposited using plasma sputtering of a Hf metal layer followed by a post-deposition oxidation. A stoichiometric hafnium dioxide layer with a relative permittivity of 22, and a bandgap around ~5.2 eV were achievable. The formation of the interfacial layer between the $HfO_2$ and the silicon, while not eliminated, has been controlled through thermal post-processing.

## ACKNOWLEDGMENTS

Authors are grateful to the National Sciences and Engineering Council of Canada (NSERC), the Canadian Foundation for Innovation, Manitoba Innovation Fund and the Canada Research Chairs (CRC) program. We would also like to thank Dr. A. Knights and his team at McMaster University for their help in TEM, and Dr. R. Wallace of University of Texas for his assistance with the XPS results.

## REFERENCES

1. J. Robertson, "High Dielectric Constant Oxides", *Eur. Phys. J. Appl. Phys.*, 28, (2004) 265-291.
2. S. Sayan, E. Garfunkel, S. Suzer, "Soft x-ray photoemission studies of the $HfO_2$/$SiO_2$/Si system", Applied Physics Letters, 80 (2002) 2135-2137.
3. (CRC Handbook) *Chemistry and Physics,* 83rd ed., David R. Lide, CRC Press, (2002-2003), **9**-52–**9**-57.
4. U. Gosele, T. Y. Tan, "Oxygen Diffusion and Thermal DonorFormation in Silicon", *Appl. Phys. A.*, 28, (1982) 79-92.
5. C. Evans. (2007, May 7). "Hafnium Oxide ($HfO_2$) Composition and Stoichiometry". *EAG labs*. [Online]. Available: http://www.eaglabs.com/files/appnotes/AN400.pdf.
6. S. Rudenja, A. Minko, and D. A. Buchanan, "Low temperature deposition of stoichiometric $HfO_2$ on silicon: Analysis and Quantification of the $HfO_2$ /Si interface from electrical and XPS measurements" *Appl. Surf. Sci.*, 257, (2010) 17-21.
7. M. F. Al-Kuhaili, "Optical properties of Hafnium oxide thin films and their application in energy-efficient windows" *Opt. Mater.,* 27, (2004) 383-387.

**Mater. Res. Soc. Symp. Proc. Vol. 1394 © 2012 Materials Research Society**
**DOI: 10.1557/opl.2012.803**

## Dependence of Annealing Temperature on the Conductivity Changes of ZnO and MgZnO Nanoparticle Thin Films from Annealing in a Hydrogen Atmosphere at Mild Temperatures

Christine Berven, Lorena Sanchez, Sirisha Chava, Hannah Marie Young, Joseph Dick, John L. Morrison, Jesse Huso, Leah Bergman

Department of Physics, University of Idaho, Moscow, ID 83844-0903, USA

### ABSTRACT

We report apparent robust doping of ZnO and $Mg_xZn_{1-x}O$ (x ~20%) nanoparticle films by annealing in hydrogen gas. The annealing was done at sequentially higher temperatures from about 20 °C to 140 °C. The effect of the annealing was determined by comparing current-voltage measurements of the samples at room-temperature and in vacuum after each annealing cycle. The nanoparticles were grown using an aqueous solution and heating process that created thin-films of ZnO or MgZnO nanoparticles with diameters of about 30 nm. When exposed to hydrogen gas at room-temperature or after annealing at temperatures up to about 100 °C, no measureable changes to the room-temperature vacuum conductivity of the films was observed. However, when the samples were annealed at temperatures above 100 °C, an appreciable robust increase in the room-temperature conductance in vacuum occurred. Annealing at the maximum temperature (~135-140 °C) resulted in about a factor of about twenty increase in the conductivity. Furthermore, the ratio of the conductance of the ZnO and MgZnO nanoparticle films while being annealed to their conductance at room-temperature were found to increase and then decrease for increasing annealing temperatures. Maximum changes of about five-fold and seven-fold for the MgZnO and ZnO samples, respectively, were found to occur at temperatures just below the annealing temperature threshold for the onset of the robust hydrogen gas doping. Comparisons of these results to other work on bulk ZnO and MgZnO films and reasons for this behavior will be discussed.

### INTRODUCTION

ZnO and MgZnO having a large band gap of 3.23 eV and 3.6 eV, have a wide range of potential applications such as in solar cells, gas sensors, chemical sensors, electrical devices, and luminescent devices. They have also been used for low voltage and short-wavelength electro-optical device applications due to their large exciton binding energies of ~60 meV.[1] ZnO and MgZnO nanoparticles have also recently been of great interest for the detection of pollutants and toxic gases.[2]

The motivation for this work was to understand how the electrical properties of ZnO and MgZnO semiconducting nanoparticles can be modified through doping by exposure to hydrogen gas. To accomplish this, changes in the electrical properties of films of ZnO and MgZnO nanoparticles were studied as a function of annealing temperature in a hydrogen ambient. Similar work was done by this research team on ZnO nanowires, where a significant response to the presence of $H_2$ gas at elevated temperatures was found.[3]

## EXPERIMENTAL DETAILS

The ZnO and MgZnO nanoparticles were synthesized by mixing magnesium acetate and zinc acetate in an aqueous solution followed by heating and annealing in a tube furnace. The resultant average crystalline size was ~30 nm. Evidence of achieving an alloy with a Mg composition of ~20% was observed as a band-shift of about 200 meV in the photoluminescence of the nanoparticle films.[4,5]

To contact the nanoparticle film, a simple non–lithographic technique was used.[6] The substrate on which the nanoparticles were grown was placed on a piece of glass (2.5 cm x 2.5 cm) and the sample was secured with silver paint. Two gold wires 25 μm diameter were placed in parallel with a separation of about 5 mm across the surface of the nanoparticle film. To maintain good electrical contact, the nanoparticles and the gold wires were compressed together with a light spring between the first and a second piece of glass.

The current-voltage (I-V) measurements were taken using a Keithley 236 Source Measure Unit ($Z_{input} \sim 10^{14} \Omega$) using voltage sweeps of 0 to ±5 V and step interval of 0.05 V with a ten-second dwell time per data point. These parameters were found to be slow enough so that the system was in electrical equilibrium and the voltage step was close enough to resolve any important features in the I-V characteristics.

The experiments were performed in an environmental chamber (volume ~ 2 ℓ) equipped with gas supply and a vacuum pump. The chamber could be pressurized to 5 psi above atmosphere and evacuated to ~5 mTorr.[7] The temperature of the sample chuck could be increased from room temperature (RT) to 160 ˚C by using heating tape wrapped around a copper extension of the thermal finger that passed through the bottom of the environmental chamber. Temperature stability to within ±2 ˚C over several hours was achieved.

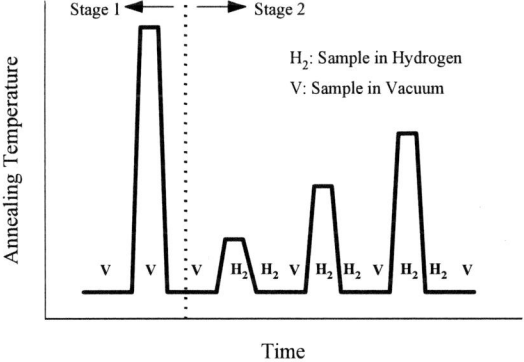

**Figure 1.** Summary of annealing history and gas exposure. At each step in the annealing cycle the system was allowed to come to thermal and electrical equilibrium before proceeding.

A cyclic annealing process was used where the sample was heated and cooled to RT in an $H_2$ atmosphere (~3 psig) at sequentially higher annealing temperatures ($T_{anneal}$) followed by evacuation. The relative temperatures and atmospheres at each step is shown in figure 1. An initial annealing in vacuum up to a temperature of 160 °C (Stage 1) was performed to drive off any surface adsorbates. Between each annealing step in $H_2$, and after the temperature was returned to RT, the chamber was evacuated (Stage 2). At each stage and before the next annealing step was performed, the I-V characteristics were recorded to verify if the system had reached equilibrium by noting no detectable change in consecutive I-V characteristics. The effectiveness of annealing to dope the nanoparticles was determined by recording how the RT current at a bias of 5 V had changed since the initial annealing in vacuum.

## DISCUSSION

### Experimental results

In the first stage of the annealing process, while in vacuum and at the maximum temperature of 160 °C, the current passing through the ZnO and MgZnO nanoparticles increased by about a factor of 15 and 20, respectively, relative to when at room temperature as would be expected for semiconducting materials. After cooling back to RT and the system reached equilibrium, the I-V plots were found to be almost identical to the plots from before the initial heating. This demonstrated that under vacuum, the system was completely reversible and refreshable and that any absorbates present from before the initial heating had no effect on the electrical properties of the nanoparticles. All I-V characteristics were found to be piece-wise ohmic and had the same proportional changes in conductivity about a bias of 0 V.

In the second stage, the nanoparticle films were exposed to $H_2$ gas at successively higher annealing temperatures. At the beginning of each annealing step when the nanoparticle films were exposed to $H_2$ at RT, no detectable change in the conductivity was measured. After annealing in $H_2$ at temperatures up to about 100 °C, the RT, vacuum conductivity of the films was found to be unchanged. However, after annealing at higher temperatures in $H_2$ the RT conductivity in vacuum increased dramatically. Figure 2 shows the overlaid plots of the currents at 5 V under vacuum at RT after each annealing step for both the ZnO and MgZnO showing that the threshold annealing temperature and the degree of increase were similar for both samples.

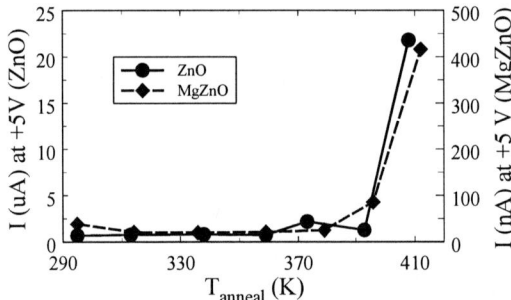

**Figure 2**. Currents in films of ZnO (●) and MgZnO (◆) nanoparticles at + 5 V showing significant rise in currents at the same threshold temperature of about 120 ˚C.

The ratio of the conductivity of the films while at the annealing temperature to the conductivity of the films after being returned to RT was also found to depend on the annealing temperature, as shown in figure 3. The ratio of the current while at $T_{anneal}$ to the current at RT after annealing steadily increased and then decreased; the peak being at about 10 to 20 K below the onset of the robust conductivity increase illustrated in figure 2. There is a small variation in the ratios of the RT conductivity with and without $H_2$ present. The maximum variation of the ratios of the conductivity at RT shows no systematic dependence on the annealing temperature. For the ZnO sample, this variation was no more than about 35% and for the MgZnO sample, this variation was no more than about 17%. These variations are quite small compared to the relative increase in the currents while at the annealing temperature of 710% for the ZnO sample and 450% for the MgZnO sample. The differences in the values of the conductance ratios at $T_{anneal}$ during the annealing is an artifact of the variation in the ratio of the RT conductance with $H_2$ and in vacuum. All three ratios were included for completeness.

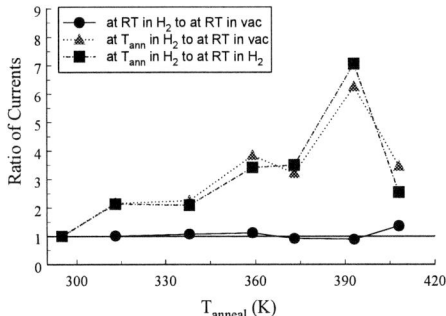 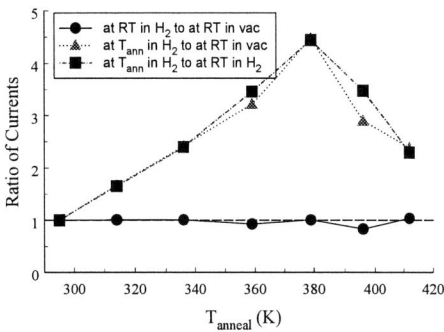

**Figure 3**. The ratio of currents as a function of annealing temperature and $H_2$ exposure for ZnO (left) and MgZnO (right) nanoparticle films.

### **Discussion of experimental results**

While at RT, the conductivity of both the nanoparticle films was found to not change while exposed to hydrogen gas. However, after annealing at sufficiently high temperatures, a robust (~days) increase in the RT, vacuum conductivity was observed for both samples (figure 2). These results are qualitatively consistent with previous work on the interaction of ZnO with hydrogen gas where hydrogen has been identified as being able to act as a donor.[3,8–11] The effect of the annealing process was similar for both samples. For each, the threshold temperature was about 120 ˚C and the relative increase in conductivity was by about a factor of twenty. The resemblance of the responses of each sample suggests that the mechanism responsible would be similar for both with the most likely mechanisms for the doping effect falling into two categories: adsorption or inclusion of the H atoms as interstitial or substitutional donors.

Gas molecules adsorbed onto a metal-oxide surface can have a strong chemical bond that will either withdraw or donate electrons to the metal-oxide.[12] This mechanism is normally reversible when the gas is removed making ZnO a candidate for use for gas sensing.[3,13–15] In the case of ZnO and MgZnO nanoparticles, it has been well established that the behavior of negatively charged chemisorbed oxygen on the grain boundaries of metal oxides determines the sensitivity to $H_2$ gas at higher temperatures.[16] Although, ZnO and MgZnO do interact with gasses through adsorption, this is a reversible process and therefore another mechanism is needed that can provide for a stable (or at least a metastable) state in which the conductivity of the ZnO and MgZnO nanoparticles can be increased.

In general, hydrogen is known as a ubiquitous and amphoteric impurity in common semiconductors. However, in ZnO hydrogen has been found to act as a shallow donor.[4] Theoretical predictions suggest that hydrogen can become an interstitial impurity in ZnO with a hydrogen atom donating its electron to the crystal where the most stable site for the $H^+$ was found to be between Zn and O atoms. In addition to single $H^+$ ions being incorporated into the lattice, a second mechanism by which a hydrogen atom combines with an oxygen vacancy has also been suggested as a way hydrogen can behave as a shallow donor.[11]

Experimental support for a stable or meta-stable doping effect by hydrogen introduced during crystal growth was first reported in the 1950's.[9]. Later, post-growth doping of ZnO has been reported by several groups for both ZnO and MgZnO.[6,10,17] In these studies, the effect of doping was found to necessitate annealing the samples at temperatures of at least 300 °C or up to 700-800 °C.

In contrast to these four studies which were carried out on bulk ZnO samples, the results reported here were carried out on thin films of nanoparticles of ZnO and MgZnO. Additionally, the threshold temperature for doping with hydrogen on the ZnO and MgZnO nanoparticles was found to be little above 100 °C. Since the materials under study are nominally similar to those where the threshold temperature for doping was higher, the difference in the morphology of the samples suggests it as an explanation for the difference in the threshold temperature for doping.

An expected consequence of the much larger surface to volume ratio of the nanoparticle films compared to the bulk samples would be to affect the speed or thoroughness of the absorption of the hydrogen. Although the energies of donor states would not be expected to change as a function of crystal size, the relatively low temperatures for the threshold of robust doping suggests otherwise. It has been reported that for films less than 200 nm thick, compressive strain in the lattice can exist.[18] Having been synthesized using a wet-chemical process, the nanoparticle crystal structure would therefore not be affected by strain with their growth substrate. The differences in the amount of strain could be a possible explanation for the different energy thresholds for the hydrogen incorporation.

The correlation as a function of temperature between the increase in total conductivity (figure 2) and the relative conductivity while at the annealing temperatures (figure 3) provides possible clues as to the doping mechanism. The increase in the RT, vacuum conductivity after successive annealing steps would be consistent with a greater number of H-donor atoms being incorporated into the crystal to act as donors. It would be expected that the number of occupied donor states would eventually saturate as the doping process continued. The observed increase in the ratios could be a reflection of the increasing ease that H can be incorporated as a donor as the annealing temperatures were increased. That there was a maximum to this behavior could be evidence that, although the conditions for incorporation of the H atoms was favorable (i.e. high temperatures), the number of available unoccupied sites could be decreasing. This would indicate a saturation of the doping sites was occurring.

Unfortunately, the work reported here was not able to include structural studies of the un-doped and doped nanoparticles nor to investigate the importance of particle size to the doping process. Pursuing such investigations, including extending the experiments to higher temperatures, could shed light on the doping mechanisms and the importance of the morphology and size of the nanoparticles to the threshold temperatures for hydrogen doping in both ZnO and MgZnO nanoparticles.

**CONCLUSION**

The ability to dope films of nanoparticles of ZnO and MgZnO synthesized from a wet-chemical technique by exposure to hydrogen was explored. Initial exposure to hydrogen gas at room temperature found the nanoparticle films insensitive to its presence. By annealing at successively higher temperatures, a threshold temperature for a dramatic increase in RT conductivity was found. This behavior was found to happen at about the same temperature of about 390 K for both the ZnO and MgZnO nanoparticles films. The conductivity increase was found to be robust, lasting for several days while stored in the dark at room temperature and under vacuum. The similarity in the behavior of the two samples suggests that the doping mechanism for both materials is the same. That the threshold temperature for these films is lower than that reported by previous studies on bulk ZnO suggests the morphology and/or the method of synthesis of the nanoparticles plays an important role in the doping mechanisms. The increase and then decrease of the ratio of the conductivity of the films while in an $H_2$ atmosphere at the annealing temperature to afterwards at RT could be evidence that the number of available donor sites for irreversible or metastable incorporation of hydrogen donor atoms could be close to saturation. This robust effect could lead to a simple way to modify the conductivity of ZnO and MgZnO nanomaterials, increasing their utility for very many electrical applications.

## ACKNOWLEDGMENTS

This work was supported by the Department of Energy Award DE-FG02-04ER46142 and National Science Foundation Award PHY-0754360.

## REFERENCES

1. H. Pan, J. Luo, H. Sun, Y. Feng, C. Poh, and J. Lin, *Nanotech.* **17**, 2963 (2006).
2. H. Tang, M. Yan, X. Ma, H. Zhang, M. Wang, and D. Yang, *Sens. Actuators, B* **113**, 324 (2006).
3. D. Zhang, S. Chava, C. Berven, S.K. Lee, R. Devitt, and V. Katkanant, *Applied Physics A: Materials Science & Processing* **100**, 145 (2010).
4. L. Bergman, J. Morrison, X. Chen, J. Huso, and H. Hoeck, *Appl. Phys. Lett.* **88**, 023103 (2006).
5. S. Chava, H.M. young, L. Sanchez, J. Dick, J.L. Morrison, J. Huso, L. Bergman, and C. Berven, in *Nanotechnology (IEEE-NANO), 2011 11th IEEE Conference on 15-18 Aug. 2011* (IEEE, Portland Marriott, Portland, Oregon, USA, 2011), pp. 1025–1029.
6. W. Liu, B. Yao, Y. Li, B. Li, C. Zheng, B. Zhang, C. Shan, Z. Zhang, J. Zhang, and D. Shen, *Appl. Surf. Sci.* **255**, 6745 (2009).
7. H.H. Hsu, H.P. Wang, C.Y. Chen, C.J.G. Jou, and Y.-L. Wei, *J. Electron. Spectrosc. Relat. Phenom.* **156-158**, 344 (2007).
8. L. Schmidt-Mende and J.L. MacManus-Driscoll, *Materials Today* **10**, 40 (2007).
9. D.G. Thomas and J.J. Lander, *J. Chem. Phys.* **25**, 1136 (1956).
10. J. Sann, A. Hofstaetter, D. Pfisterer, J. Stehr, and B.K. Meyer, *Phys. Status Solidi C* **3**, 952 (2006).
11. C.G. Van de Walle, *Phys. Rev. Lett.* **85**, 1012 (2000).
12. M. Heinrich, C. Domke, P. Ebert, and K. Urban, *Phys. Rev. B* **53**, 10894 (1996).
13. Z. Fan and J.G. Lu, *Appl. Phys. Lett.* **86**, 123510 (2005).
14. E. Comini, G. Faglia, M. Ferroni, and G. Sberveglieri, *Appl. Phys. A* **88**, 45 (2007).
15. S. Sathanantha, V.P. Dravid, and S.-W. Fan, *Nanoscape* **6**, 6 (2009).
16. F. Lin, Y. Takao, Y. Shimizu, and M. Egashira, *Sens. Actuators, B* **25**, 843 (1995).
17. M.D. McCluskey and S.J. Jokela, in *Proceedings of the NATO Advanced Workshop on Zinc Oxide* (2005), pp. 125–132.
18. S.H. Park, T. Hanada, D.C. Oh, T. Minegishi, H. Goto, G. Fujimoto, J.S. Park, I.H. Im, J.H. Chang, M.W. Cho, T. Yao, and K. Inaba, *Appl. Phys. Lett.* **91**, 231904 (2007).

Mater. Res. Soc. Symp. Proc. Vol. 1394 © 2012 Materials Research Society
DOI: 10.1557/opl.2012.825

## Fabrication of Titanium Oxide Film with high crystallinity by the New Electrochemical Techniques

Hiroki Ishizaki[1] and Seishiro Ito[2]

[1] Department of Electronic System Engineering, Tokyo University of Science Suwa,
5000-1 Toyohira, Chino-shi, Nagano 391-0292, Japan
[2] Faculty of Science and Engineering, Kinki University,
4-1 Kowakae 3-chome, Higashiosaka, Osaka 577-8502, Japan

### ABSTRACT

$TiO_2$ films with poly crystal were electrodeposited on conductive substrate (NESA glass, approximately $10 \, \Omega/\square$, Asahi glass Co. Ltd.) from the titanium potassium oxalate dehydrate aqueous solution containing hydroxylamine adjusted pH 9 with KOH aq. at 333k. The peak corresponded to $Ti^{3+}$ion into these $TiO_2$ film was not observed by using X-ray photoelectron spectroscopy (XPS). The photocatalysis of $TiO_2$ film increased with a decrease of cathodic potential. In particular, $TiO_2$ film obtained at cathodic potential of -1.3V, had the higher photocatalysis than that of other potential.

### INTRODUCTION

Recently, titanium oxide films are paid much attention for many applications such as photocatalysis, chemical sensor, ferroelectrical devices, opto-electrical devices and solar cells [1, 2], because of its ferroelectrical, photocatalytic and optical properties. In order to develop the dye-sensitized solar cell devices with high-performance, the photocatalytic property and crystallinity of titanium oxide film need be improved [3, 4]. Thus, we will suggest that titanium oxide film with high crystallinity will be obtained on the conductive substrate at the low temperature by electrochemical techniques.

In particular, the electrochemical deposition techniques presents several advantages: (1) relatively uniform film can be obtained on substrates with melting point lower than 373K, (2) the thickness and morphology of films can be controlled by electrochemical parameters, (3) the deposition rate is relatively high, (4) the equipment is not expensive, and (5) the process is less hazardous and environmentally friendly [5].

In this paper, we reported the detail results for the fabrication of poly crystalline titanium oxide films on conductive substrate (NESA glass) from the titanium ion aqueous solution containing a complex agent and a hydroxylamine at pH9 by electrodeposition without the heat treatment.

### EXPERIMENT

The $TiO_2$ film is grown by cathodic potential ranging of -1.0V to -1.3V referred to Ag/AgCl electrode in 0.05 mol/L titanium potassium oxalate dehydrate ($K_2[TiO(C_2O_4)_2]$) aqueous solution containing 0.5mol/L hydroxylamine ($NH_2OH$). The solutions prepared with distilled water and reagent grade chemicals. And then this solution with colorless and transparent was obtained by the adjusting pH9 with KOH aq. The conductive substrate (NESA glass) is used as the cathode. Prior to deposition, the glass substrate (10x20x1.1mm) was rinsed in acetone and anodically polarized in a 1.0mol/L NaOH aqueous solution and was then active anode. The Pt/Ti sheet (99.99%purity) is used as active anode. The Ag/AgCl electrode in saturated KCl aqueous solution is used as the reference electrode. And the thickness of $TiO_2$ film was controlled by the electric charge. The preparation of $TiO_2$ film with the thickness of 50µm is carried out potentiostatically using a potentiostat (Hokuto Denko, HABF501) without stirring until the electric charge of 10 coulomb $cm^{-2}$.

X-ray diffraction measurements are performed with Rint 2000 using monochromated Cu Kα radiation operated at 30kV and 16mA. The electron spectrum of titanium and oxygen in $TiO_2$ film were recorded with X-ray photoelectron (XPS, Shimadzu 830). The cross-sectioned morphology and surface morphology of $TiO_2$ films were observed by using a scanning electron microscopy (N-SEMEDX, Hitachi S2200). The optical properties of these thin films are measured by using UV-VIS-NIR scanning spectrophotometer (UV-VIS-NIR, Shimadzu, MPC3100). The photocatalytic activities of the samples for $CH_3CHO$ oxidation were examined. A 508-ppm standard $CH_3CHO$ gas ($CH_3CHO/N_2$) was introduced into a reaction chamber (3.3 L), and diluted with air so that the initial concentration was controlled within the 205 ± 10 ppm range. After the adsorption equilibrium of $CH_3CHO$ had been achieved in the dark, front-face irradiation ($\lambda > 300$ nm, I(at the range of 320nm from 400nm = 2.9mW $cm^{-2}$) of the sample (apparent $TiO_2$ surface area = 2 $cm^2$) was started using a 300 W Xe lamp (Wacom, model XDS-301S) at room temperature. The concentrations of $CH_3CHO$ were determined as a function of illumination time by gas chromatography (Shimadzu GC-9A; f.i.d. column Shimadzu SHINCARBONA ($\varphi$3 mm × 3 m)). The carrier gas was $N_2$ (0.5 kg $cm^{-2}$) at an injection temperature of 70°C and a column temperature of 70°C.

**DISCUSSION**

Figure 1 shows the XPS spectra of $TiO_2$ films obtained at the cathodic potential of -1.0V, -1.2V and −1.3V, respectively. Before sputter cleaning the surface of $TiO_2$ films, a carbon 1s peak is observed at the vicinity of 284.5eV, regardless of cathodic potential. The presents of this peak is related to the organics surface contamination corresponding to that of C1s in hydrocarbon. The XPS spectra of the $TiO_2$ films are obtained after a sputter cleaning for 1min in order to remove the surface contaminants. These C1s peaks aren't observed in the $TiO_2$ films after the sputter cleaning. The peaks observed at about 459eV and about 880ev are identified to Ti 2p3/2 and Ti LMM for $Ti^{4+}$ in $TiO_2$ envelope, regardless of cathodic potential [6]. And O1s and OKLL peaks for $O^{2-}$ of $TiO_2$ are observed at about 530eV and about 780eV, respectively [7]. No other contaminants are observed at the $TiO_2$ films. And the structural properties of these $TiO_2$ film were evaluated by the x ray diffraction measurements with using monochromatic Cu Kα radiation operated at 45kV and 40mA. All diffraction lines are assigned to those of $SnO_2$ and $TiO_2$ with anatase structure, regardless of cathodic potential. Any diffraction lines attributed to other titanium compounds cannot be observed.

Figure 2-a) shows the effect of cathodic potential on the Ti 2p XPS spectra of titanium dioxide film electrochemically obtained from a titanium potassium oxalate dehydrate aqueous solution containing a hydroxylamine kept at 333K and pH9. The peaks observed at about 465.0eV for Ti 2p1/2 and about 459.3eV for Ti2p3/2, regardless of the cathodic potential [8, 9]. These binding energies are much closed to the values of the $Ti^{4+}$ valence states of $TiO_2$ with anatase structure. We know that the temperature above 1073K is required to converted anatase structure to the rutile structure. The standard free energy of anatase structure was lower than that of rutile structure at temperature above room temperature, referring of $TiO_2$ phase diagram [10]. Closed $TiO_2$ formation reaction to the equilibrium reaction, $TiO_2$ with anatase structure will be obtained on the substrate, regardless of the cathodic potential. Figure 2-(b) shows the O1s XPS spectra of $TiO_2$ films. The peaks of O1s spectrum is observed at the vicinity of 530eV corresponding to that for O1s of $TiO_2$ envelope, regardless of the cathodic potential [11]. M. Shirkhanzedeh reposted that O1s spectrum for a titanium hydroxide was observed at about 534eV [12]. However, O1s peak for the titanium hydroxide was not observed at the binding energy of about 534eV, regardless of cathodic potential.

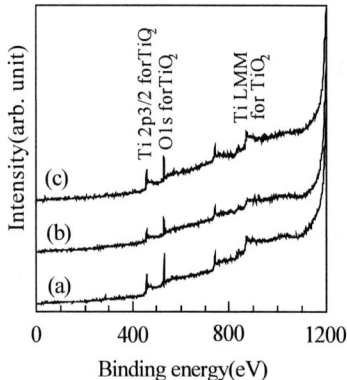

Figure 1. The XPS spectra of $TiO_2$ films obtained at the cathodic potential of (a)-1.0V, (b)-1.2V and (c)–1.3V, respectively.

The authors will propose a tentative electrochemical growth reaction for $TiO_2$ films, as described by the following reactions.

$$K_2TiO(C_2O_4)_2 \rightarrow 2K^+ + TiO(C_2O_4)_2^{2-} \tag{1}$$

$$NH_2OH + 2H_2O + 2e^- \rightarrow NH_4^+ + 2OH^- \tag{2}$$

$$TiO(C_2O_4)_2^{2-} + 2OH^- \rightarrow TiO(OH)_2 + 2C_2O_4^{2-} \tag{3}$$

$$TiO(OH)_2 \rightarrow TiO_2 + H_2O \tag{4}$$

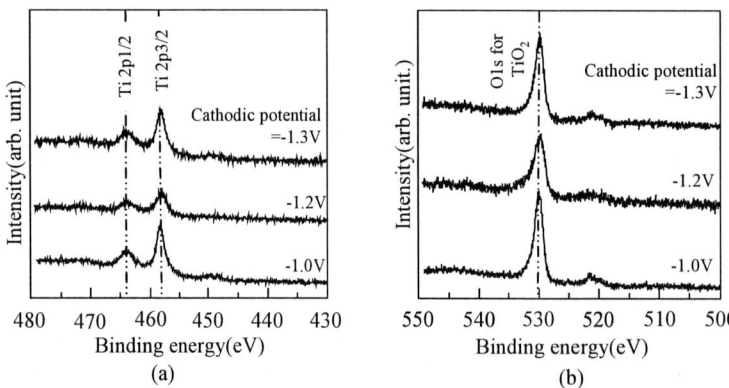

Figure 2. The effect of cathodic potential on the Ti 2p and O1s XPS spectra of titanium dioxide film electrochemically obtained on NESA glass.
(a) Ti 2p spectra and (b) O1s spectra

The reduction reaction of hydroxylamine and existence of $TiO^{2+}$ ion into the electrolyte play the important roles in the formation of $TiO_2$ film. The reaction rate of hydroxylamine will strongly be influenced on the cathodic potential, according to equation (2). Thus, $TiO_2$ films with polycrystalline structure were obtained on conductive substrates by electrochemical deposition.

Figure 3. the surface morphology of 50µm-thick $TiO_2$ films

Figure 3 shows the surface morphology of 50µm-thick $TiO_2$ films electrochemically obtained from a 0.05mol/L titanium potassium oxalate dehydrate aqueous solution with 0.5mol/L hydroxylamine at 333K and pH9. $TiO_2$ films were composed of aggregates of tetragonal grains with the average grain size of about 10µm, regardless of cathodic potential. These films had no smooth surface and many defects such as pores, regardless of cathodic potential. And the many dendrites of $TiO_2$ exist in the surface of $TiO_2$ film with increasing in the cathodic potential. Since the current density rapidly increases with an increase in the cathodic potential, it indicates that the increase in current density at one part of the surface of $TiO_2$ film will give the growth of the dendrite of $TiO_2$.

The transmission data and reflection data measurements of near-normal incidence light at the wavelength ranging from 300nm to 800nm are performed by the using UV-VIS-NIR scanning spectrophotometer at the temperature. The absorption coefficient of the films is calculated by the following the expression [13].

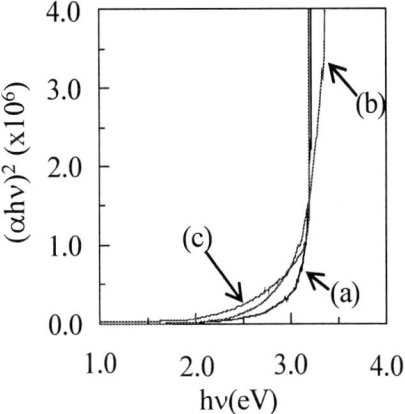

Figure 4 The $(\alpha h\nu)^2$ plots as a function of photon energy (hv)
(a)Cathodic potential of -1.0V, (b)-1.2V, (c)-1.3V

$$T=(1-R)^2 exp(-\alpha d)=(1-R)^2 exp(-4\pi kd/\lambda)^2 \qquad (5)$$

Where R is the reflectance, $\alpha$, is the absorption function, k is the extinction coefficient, d is the thickness of film, $\lambda$ is the wavelength of the incident electromagnetic wave, and $\alpha = 4\pi k/\lambda$. For the investigation of the optical bandgap energy of $TiO_2$ film with the anatase structure, Tang et al reported that $TiO_2$ film with anatase structure had the direct bandgap [14]. Thus, the bandgap energy of $TiO_2$ film is evaluated by the extrapolating the $(\alpha h\nu)^2$-photo energy (E=h$\nu$) plots. The $(\alpha h\nu)^2$ plots as a function of photon energy (h$\nu$) is shown in figure 4. $TiO_2$ films have the optical bandgap energy of about 3.2eV, regardless of the cathodic potential. Other authors reported that the optical bandgap energy of $TiO_2$ film with a rutile structure was lower than that of $TiO_2$ film with an anatase structure [14-16]. In particular, $TiO_2$ film had the optical bandgap energy of 3.0eV and 3.2eV for rutile structure [15] and anatase structure [16], respectively. And the optical bandgap energy strongly depended on the crystal structure. Thus, the $TiO_2$ film with the anatase structure will be electrochemically grown with the decrease in the cathodic potential referring of optical property, figure 2 and figure 3.

   Figure 5 shows the photocatalytic decomposition of $CH_3CHO$ for $TiO_2$ film electrochemically grown at cathodic potential of (a) –1.0V, (b) –1.2V and (c) –1.3V, respectively. The apparent rate constants (k) were calculated from the slopes of the straight lines in the plots of $\ln(C_0/C)$ vs the illumination time. $C_0$ and C denote the concentrations of $CH_3CHO$ at t = 0 and t = t, respectively. For the cathodic potential of –1.0V, the straight line is observed at the illumination time below 0.6h. For the cathodic potential of –1.2V, the straight line is observed at the illumination time below 1.2h. For the cathodic potential of –1.3V, the straight line appears at the illumination time below 1.5h. The apparent rate constants (k) are $0.0929h^{-1}$, $0.0536h^{-1}$ and $0.0501h^{-1}$ for the cathodic potentials of –1.3V, -1.2V and –1.0V, respectively. The apparent rate constants (k) increase with an increase in the cathodic potential. Figure 3 indicates that the grain size of $TiO_2$ film decreases with the increasing in the cathodic potential. Thus, the apparent rate constants will strongly depend on the grain size of $TiO_2$ film. It indicates that the surface

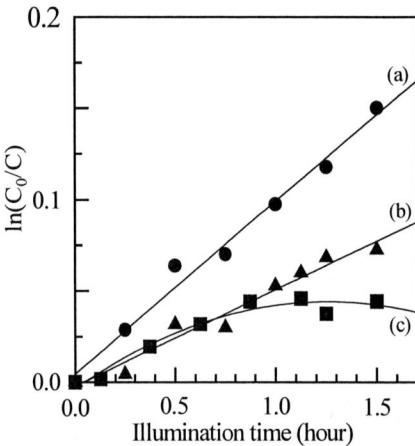

Figure 5 The photocatalytic decomposition of $CH_3CHO$ for $TiO_2$ film electrochemically grown on NESA glass. (a) Cathodic potential of -1.3V, (b) -1.2V and (c) -1.0V

morphology and the grain size of $TiO_2$ film will play an important rule to fabricate $TiO_2$ film with the high photocatalytic activity.

## CONCLUSIONS

The pours and the polycrystalline $TiO_2$ films are prepared on the conductive glasses (NESA glass,) from a 0.05mol/L titanium potassium oxalate dehydrate aqueous solution containing the 0.5 mol/L hydroxylamine kept at 333K by electrochemical deposition without the heat treatment. These $TiO_2$ films are composed of aggregates of tetragonal grains, regardless of cathodic potential. The grain size of $TiO_2$ films increases with an increase in cathodic potential. And the optical band gap energy of $TiO_2$ film decreases with an increase in the cathodic potential. Since the rutile structure changes from the anatase structure by the increasing in cathodic potential, the optical band gap energy of $TiO_2$ film would be narrow with the increase in the cathodic potential. And these photocatalytic activities of $TiO_2$ films increase with an increase in cathodic potential. Thus, $TiO_2$ film with photocatalytic activity will be grown on the conductive glass substrate (NESA glass) by electrochemical reaction without the heat treatment. Since $TiO_2$ compound are ceramic material extensively used in the photocatalytic and corrosion engineering industry field, this investigation will give the development of the photocatalytic and corrosion engineering field.

## ACKNOWLEDGMENTS

This work was supported in part by Adaptable and Seamless Technology Transfer Program through target-driven R&D (A-STEP) AS231Z03075B

## REFERENCES

1. P. Lobl, M. Huppertz and D. Mergel, *Thin Sold Films*, **251**, 72 (1994)
2. H. K. Ha, M. Yoshimoto, H. Koinuma, B. K. Moon and H. Ishiwara, *Appl. Phys. Lett.*, **68**, 2965 (1996)
3. H. Sankur and W. Gunning, *J. Appl. Phys.*, **66**, 4747 (1989)
4. M. Grätzel, Nature, **414**, 338 (2001)
5. A. Hattori, Y. Tokihisa, H. Tada, N. Tohge, S. Ito, K. Hongo, R. Shiratsuchi and G. Nogami, *J. Sol-Gel Sci. Tech.*, 22, 53 (2001)
6. J. F. Moulder, W. F. Stickle, P. E.Sobol and K. D. Bomben, *"Handbook of X-ray Photoelectron Spectroscopy"*, J. Chastain, Editor, Perkin-Elmer Corporation Physical Electronics Division, America (1992). p.72,
7. J. F. Moulder, W. F. Stickle, P. E.Sobol and K. D. Bomben, *"Handbook of X-ray Photoelectron Spectroscopy"*, J. Chastain, Editor, , Perkin-Elmer Corporation Physical Electronics Division, America (1992). p.44
8. N. Noel and P. N. Anantharaman, *J. Electroanal. Chem.*, **191**, 127 (1985).
9. H. Y. Lee, Y. H. Park and K. H. Ko, *Langmuir*, **16**, 7289 (2000).
10. H. Okamoto, *J. Phase Equilib.*, **22**, 515 (2001).

11. L. D. Piveteau, M. I. Girona, L. Schlapbach, P. Barboux, J. P. Boilot and B. Gasser, *J. Mater. Sci.*, **10**, 161 (1999).
12. M. Shirkhanzadeh, *J. Mater. Sci.*, **6**, 206 (1995).
13. A. R. Forouhi, I. Bloomer, *Phys. Rev. B*, **15**, 7018 (1986)
14. H. Tang, F. Lèvy, H. Berger and P. E. Schmid, *Phys. Rev. B*, **52**, 7771 (1995)
15. H. Tang, H. Berget, P. E. Schmid, F. Levy and G. Burri, *Solid State Commun.* **87**, 847 (1993)
16. J. Pascual, J. Camassel and H. Mathieu, *Phys. Rev. B*, **18**, 5606 (1978)

Mater. Res. Soc. Symp. Proc. Vol. 1394 © 2012 Materials Research Society
DOI: 10.1557/opl.2012.826

## Effects of Substrate Pre-deposition Annealing and Deposition Parameters on the Properties of RF Sputter-deposited ZnO Films

T. N. Oder, M. McMaster, A. Smith, N. Velpukonda and D. Sternagle

Department of Physics and Astronomy, Youngstown State University, Youngstown, OH 44555, USA

### ABSTRACT

Zinc Oxide thin films were deposited on sapphire substrates by radio frequency (RF) magnetron sputtering from an ultra-high purity ZnO solid target. The ZnO films were deposited on sapphire substrates heated in oxygen and/or in vacuum prior to deposition. Additional parameters investigated included the substrate temperature varied from 25 °C to 600 °C, the deposition gas pressure varied from 5 mTorr to 40 mTorr and the gas flow rate varied from 5 to 30 standard cubic centimeter per minute (sccm). The resulting films were annealed using a rapid thermal processor in $N_2$ gas at 900 °C for 5 min. Analyses carried out using photoluminescence spectroscopy (PL) and X-ray diffraction (XRD) measurements indicate that films deposited at 300 °C using $Ar:O_2$ (1:1) had the best optical and microstructure qualities. Pre-heating the sapphire substrate in oxygen prior to deposition was found to create a smoother sapphire surface, and this produced a ZnO film with greatly improved qualities. This film had a luminescence peak at 3.362 eV with a full-width-half maximum (FWHM) value of 15.3 meV when measured at 11 K. The XRD $2\theta$-scans had peaks at 34.4° with the best FWHM value of only 0.10°. Production of high quality ZnO materials is a necessary step towards realizing highly conductive p-type doped ZnO materials which is currently a major goal in research efforts on ZnO.

### INTRODUCTION

Zinc oxide (ZnO) is a wideband gap semiconductor with a direct energy band gap of 3.37 eV at 300 K or 3.437 at 4 K [1]. It bears similarity with gallium nitride (GaN) and hence can as well be used for fabricating electronic devices for high temperature/high power applications and optoelectronic devices for blue/UV applications. ZnO however, has several advantages over GaN, making it a candidate for replacing or complimenting GaN [2]. It has a larger exciton binding energy (60 meV for ZnO, 24 meV for GaN) making it more attractive for fabricating efficient room-temperature LEDs and lasers. It has readily available native single crystal substrate, a broad chemistry for easy processing (e.g. wet etching), and can be grown at relatively low temperature, which would help in reducing the cost of processing. The larger radiation hardness of ZnO makes it attractive for fabricating electronic devices for space exploration applications. Because it is transparent, ZnO is useful in solar cell windows, transparent thin-film transistors for large area displays and MEMS wave devices. Lastly, by doping it with Mn, Fe and Cr, ZnO can be used to make spintronic devices useful for information processing, lasers and transistors. Successful development of ZnO would introduce a semiconductor material suitable for the fabrication of a wider variety of low cost electronic and optical devices that would transform the existing technology. Currently, lack of high quality p-type ZnO materials has been a major hindrance to its development. The main reasons for the p-type doping difficulty come from low solubility and self-compensation of the acceptor dopants by defects [3-5]. Native defects such as oxygen vacancies, zinc interstitials or zinc antisites were once thought to be the cause for the unintentional n-type conductivity in ZnO [6-8]. However, while recent research have cast doubt on this concept [9], these defects nevertheless play a big role as compensating centers to p-type doping [10]. Furthermore, hydrogen-related defects have been suggested to be some of the hindrances to p-type doping in ZnO, especially when H substitutes for O in ZnO and act as a shallow donor [11]. Therefore, understanding the role of native point defects (i.e. vacancies, interstitials, and antisites) and the incorporation of impurities is key toward controlling the conductivity in ZnO. The main goal of this work was to optimize the growth and processing conditions of ZnO films with reduced defects in order to obtain successful p-type dopant incorporation.

## EXPERIMENTAL DETAILS

The c-plane sapphire used as substrates were cleaned using acetone, alcohol, buffered HF acid and loaded in a vacuum chamber, pumped down to $4 \times 10^{-7}$ Torr. ZnO films were then deposited for 2 hours using RF magnetron sputtering from an ultra pure ZnO target with an RF power of 100 watts and deposition gas that consisted of ultra-high purity (UHP) Ar and $O_2$ mixed in a ration 1:1. Post deposition annealing was carried using a rapid thermal processor (RTP) in UHP $N_2$ at 900 $^{\circ}C$ for 5min to improve the film quality. Note that the process gases used here, either for deposition or annealing were of ultra-high purity quality. The first optimization involved pre-deposition substrate treatment where the sapphire substrates were annealed by heating at 500 $^{\circ}C$ for 30 min in vacuum, in $O_2$, or in vacuum followed by $O_2$. In this first optimization, the substrate temperature was kept at 500 $^{\circ}C$ during deposition while the deposition gas flow rate used was maintained at 10 standard cubic centimeter per minute (sccm) and a pressure of 10 mTorr. The optimum condition from this set was then used in subsequent investigations. The other parameters investigated were the temperature of the substrate during deposition, which was varied from 20 $^{\circ}C$ (unheated) to 600 $^{\circ}C$; the deposition gas pressure varied from 5 mTorr to 40 mTorr and the deposition gas flow rate varied from 5 to 40 sccm. The optical properties of the films were obtained from photoluminescence (PL) spectroscopy measurements that were conducted using a 325 nm He-Cd laser source. The film microstructures were investigated by X-ray diffraction measurements using the Bruker-Nonius D8 Advance Powder Diffractometer with a Cu $K_\alpha$ line of 1.54 Å. The electrical properties were investigated using Hall effect measurements. The film surface morphology was characterized using the atomic force microscopy (AFM).

## DISCUSSIONS

Figure 1(a) shows the PL spectra collected at 11 K from ZnO films deposited on sapphire substrates pre-heated in vacuum, in $O_2$, or in vacuum followed by $O_2$ in comparison to the film deposited on unheated sapphire substrate. The emission line at 3.361 ($\pm$ 0.01) eV is attributed to the $D^0X$ donor-bound excitons [12]. As can be seen in this figure, the film deposited on substrates pre-heated in oxygen had the strongest luminescence with an integrated intensity value about 15 times that of the film deposited on unheated substrate at the 3.361 eV peak. In addition, the full width at half maximum from this film was equally at the optimum value of 15.3 meV, compared to 17.4 meV for the film deposited on the unheated substrate. Figure 1 (b) is the θ-2θ

X-ray diffraction scan of the films, which show a peak at 34.4° that corresponds to the diffraction from the (0 0 0 2) plane of ZnO and indicates a strong c-axis orientation perpendicular to the surface at the sapphire substrate. The inset shows the FWHM of the peak at 34.4° which indicates that the film deposited on $O_2$-heated substrate has a minimum FWHM value of 0.10°, further validating its superior quality. The grain sizes were determined using Scherrer's formula:

$$t = \frac{C\lambda}{B\cos\theta} \tag{1}$$

where B is the FWHM (in radians), $\lambda$ is the x-ray wavelength of the Cu $K_\alpha$ (0.154 nm), $\theta$ is the Bragg diffraction angle ($2\theta = 34.4°$), and C is a correction factor, taken as 0.89.

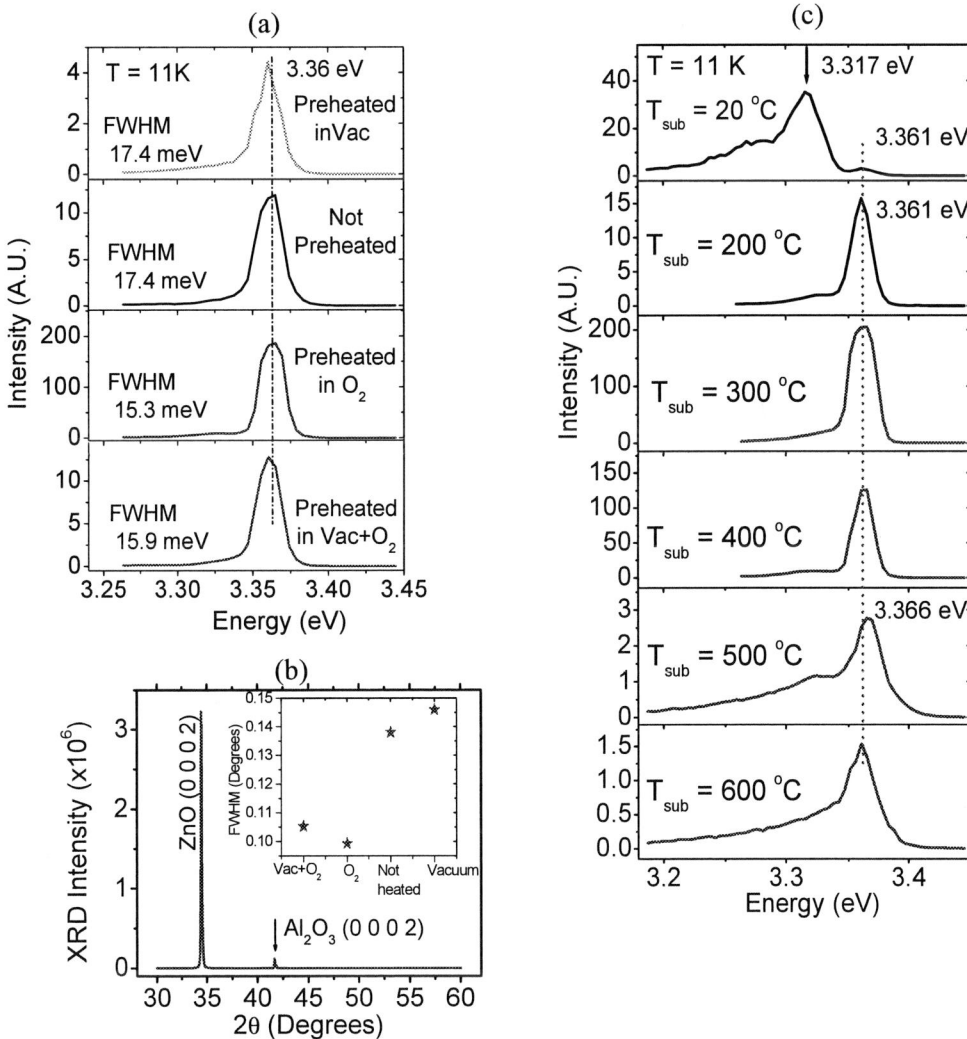

**Figure 1**. (a) Photoluminescence spectra (at 11 K) and (b) X-ray diffraction scans ($\theta$-$2\theta$) of ZnO films deposited on sapphire substrates after various pre-heating conditions. The inset in (b) shows the FWHM of the ZnO-related peak at 34.4° (c) Photoluminescence spectra (at 11 K) of films deposited on substrates maintained at different temperatures.

The estimated grain sizes obtained ranged from 56 nm to 83 nm, where the larger values indicate stress reduction and an enhancement in crystallinity of the films [13,14]. The surface morphology of the sapphire substrates without the ZnO films was evaluated after the preheating using AFM scans. It was found that pre-heating in $O_2$ leads to a much smoother surface with a root-mean square roughness of 1.6 nm, implying that the pre-heating in $O_2$ helps effectively cleans, smoothens and buffers the surface and helps improve the quality of the ZnO film deposited thereafter.

Figure 1 (c) shows the PL spectra from films that were deposited on sapphire substrates maintained at different temperatures varied from 20 °C (unheated) to 600 °C. Prior to the film deposition, the sapphire substrates were heated in $O_2$ since this was the optimum process obtained. The deposition gas pressure and flow rate used here were 10 mTorr and 15 sccm, respectively. From these spectra, it can be deduced that the films deposited on substrates held at 300 °C had the best luminescence properties. The integrated intensity at the donor-bound emission peak of 3.360 eV from this film was larger in comparison, and its FWHM value was about 19.7 meV. The spectrum from the film deposited on the substrate held at 20 °C revealed a very weak peak at 3.360 eV, but a much larger defect-related peak at 3.317 eV. Up to 300 °C, higher temperature is preferred, and this provides sufficient energy for seeding the Zn and O atoms to the substrate.

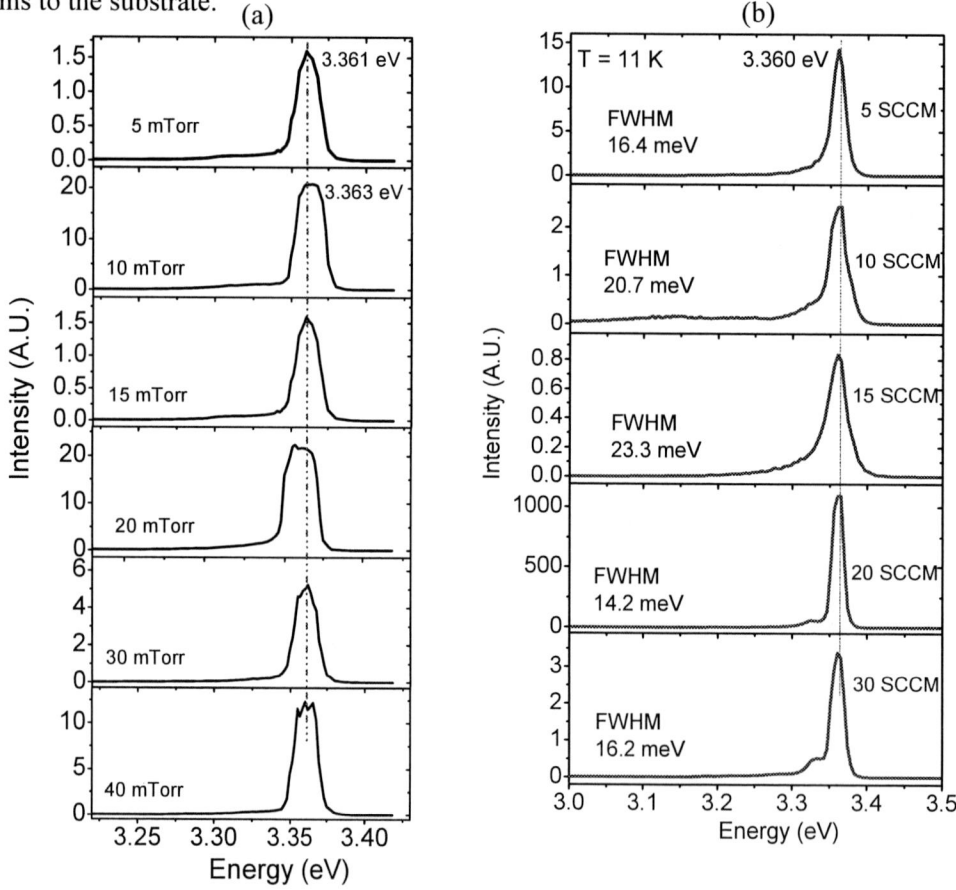

**Figure 2**. Low temperature photoluminescence spectra of ZnO films deposited on sapphire substrates (a) under different gas pressure and (b) under different gas flow rate.

Elevated substrate temperature provides kinetic energy for the sputtered Zn and O atoms to migrate to suitable crystal sites in the film growth [15,16]. The poor quality of the film when deposited while the substrate is unheated results from lack of sufficient energy for the Zn and O atoms to react with the surface atoms and form ordered growth. Above the optimum temperature, the film quality decreases perhaps due to desorption of these atoms from the surface. Other researchers have reported optimum growth temperatures of 450 °C by metal organic chemical vapor deposition [17] and 400 °C [16,18] or 250 °C [19] by pulsed laser deposition.

The PL collected at 11 K from ZnO films deposited on sapphire substrate at different pressure and different gas flow rate are shown in Figure 2 (a) and (b) respectively. Prior to the film deposition, the sapphire substrates were similarly heated in $O_2$ at 500 °C for 30 min. These spectra show that the optimum value of deposition pressure is 10 mTorr and a flow rate of 20 sccm. The film deposited at 20 mTorr had PL spectra similar to that grown at 10 mTorr, while that grown at 15 mTorr showed a poorer quality PL spectrum. Pressure and flow rate affect the kinetic energy of the sputtered atoms. Zhu, et al reported a similar pressure variation for ZnO deposited on sapphire and quartz substrates [20]. At high pressure, the atoms arrive with lower kinetic energy, implying reduced surface mobility and poorer nucleation. A lower pressure, on the other hand means the atoms have high kinetic energy to form better crystalinity. The thicknesses of the films deposited at different gas pressures were further analyzed. This was done by patterning the films using photolithography with AZ P4620 photoresist, which was subsequently hard baked to serve as etch mask. The ZnO film not covered by the photoresist was removed by wet-etching for 4-6 min in dilute hydrochloric acid (HCl: $H_2O$ = 1:60). The residual photoresist was removed by cleaning the sample in acetone, rinsing in alcohol and water and the sample was blow-dried. The thickness of the ZnO film was determined by scanning with an AFM across the etched/unetched region of the sample. Figure 3 shows a plot of the ZnO thickness variation with the gas pressure used in the deposition. Between 5 mTorr and 30 mTorr, there is a steep decrease of the film deposition rate with increasing pressure, but a much slower change between 30 mTorr and 40 mTorr. Hall-effect measurements of a ZnO film deposited at the optimum conditions indicated n-type conductivity with a concentration of 3.9 x $10^{18}$ $cm^{-3}$ and a mobility of 43 $cm^2$/V-s and a resistivity of 0.0376 Ω-cm. With the optimized growth conditions here obtained, we are currently pursuing p-type doping of the zinc oxide films using phosphorus and arsenic. Preliminary results from these efforts have yielded resistive p-type films which we hope will improve with further optimization of post-deposition annealing.

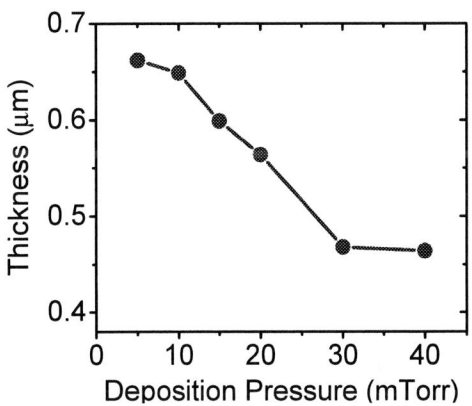

**Figure 3**. ZnO film thickness variation with the pressure of the deposition gas. The film was deposited for 2 h at an RF power of 100 watts.

## CONCLUSIONS

ZnO films were deposited using RF magnetron sputtering. Parameters changed to optimize the film quality included pre-deposition heating of the sapphire substrate in different environments, the substrate temperature during deposition, the gas flow rate and the gas pressure. The best films were obtained by depositing on sapphire substrate pre-heated in $O_2$, with a substrate temperature of 300 $^o$C, gas pressure of 10 mTorr, and gas flow rate of 20 sccm. The films produced had PL emission peaks at 3.361 ($\pm$0.001) eV, measured at 11 K, which is attributed to donor bound exciton. The XRD scan revealed a sharp diffraction peak at ~ 34.8$^o$ that is assigned to (0 0 0 2) ZnO.

## ACKNOWLEDGMENTS

The authors wish to gratefully acknowledge funds from the National Science Foundation (#DMR-1006083) which supported this work. Assistances by Dr. Matthias Zeller (Department of Chemistry, Youngstown State University) for the XRD measurements, Joshua Petrus (Department of Physics and Astronomy) for the AFM measurements and Dr. M . L. Nakarmi (Department of Physics, Brooklyn College) for the Hall effect measurements are also gratefully acknowledged.

## REFERENCES

1.  M. R. Wagner, U. Haboeck, P. Zimmer, A. Hoffmann, S. Lautenschläger, C. Neumann, J. Sann, B. K. Meyer, *Proc. of SPIE* **6474**, 64740X-1 (2007).
2.  Ü. Özgür, Ya. I. Alivov, C. Liu, A. Teke, M. A. Reshchikov, S. Doğan, V. Avrutin, S.-J. Cho, and H. Morkoç, J. Appl. Phys. **98**, 041301 (2005).
3.  C. G. Van de Walle, D. B. Laks, G.F. Neumark, and S. T. Pantelides, *Phys. Rev. B* **47**, 9425 (1993).
4.  S. B. Zhang, S.-H. Wei, and A. Zunger, *Phys. Rev. B* **63**, 075205 (2001).
5.  E.-C. Lee, Y.-S. Kim, Y.-G. Jin, and K. J. Chang, *Phys. Rev. B* **64**, 085120 (2001).
6.  K. Hoffmann and D. Hahn, *Phys. Status Solidi* a **24**, 637 (1974).
7.  A. Hausmann and B. Utsch, *Z. Phys. B* **21**, 217 (1975).
8.  K. I. Hagemark, *J. Solid State Chem.* **16**, 293 (1976).
9.  L. S. Vlasenko and G. D. Watkins, *Phys. Rev. B* **72**, 035203 (2005).
10. A. Janotti and C. G. Van deWalle, *Phys. Rev. B* **75**, 165202 (2007).
11. A. Janotti and C. G. Van de Walle, *Phys. Rev. B* **6**, 44 (2007).
12. D. C. Look, R. L. Jones, J. R. Sizelove, N. Y. Garces, N. C. Giles, and L. E. Halliburton, *Phys. Stat. Sol. a* **195**, 171 (2003).
13. M. K. Puchert, P. Y. Timbrell, R. N. Lamb, *J. Vac. Sci. Technol. A* **14**, 2220 (1996).
14. V. Gupta, A. Mansingh, *J. Appl. Phys.* **80**, 1063(1996).
15.  X. M. Fan, J. S. Lian and Z. X. Guo, *Appl. Surf. Sci.* **239**, 176 (2005).
16. A. K. Yousif and A. J. Haider, *J. Eng. Tech.* **29**(1), 58-64 (2011).
17. S. T. Tan, B. J. Chen, X. W. Sun, W. J. Fan, H. S. Kwok, X. H. Zhang and S. J. Chua, *J. Appl. Phys.* **98**, 013505 (2005).
18. S.-J. Kang, H.-H. Shin and Y.-S. Yoon, *Journal of the Korean Physical Society* **51**(1), 183-188 (2007).

19. R. C. Scott, K. DLeedy, B. Bayraktaroglu, D. C. Look,  D. J. Smith, D. Ding, X. F. Lu and Y. H. Zhang, *J. Electron. Mater.* **40**(4), 417-428 (2011).
20. S. Zhu, C.-H. Su, S. L. Lehoczky, P. Peters, M. A. George, *Journal of Crystal Growth* **211**, 106 -110 (2000).

# AUTHOR INDEX

Adams, P. ............................................. 53

Barrie, J. ............................................. 53

Belo, G. ............................................. 108

Belova, L. ............................................. 13

Bergman, L. ........................... 21, 48, 114

Berven, C. ............................................. 114

Blanchard, B. ............................................. 48

Boonchun, A. ............................................. 27

Buchanan, D. ............................................. 108

Cabarrocas, P. ............................................. 42

Charpentier, C. ............................................. 42

Chava, S. ............................................. 114

Che, H. ............................................. 48

Covington, L. ............................................. 7

Cullen, J. ............................................. 1

Dick, J. ............................................. 114

Fang, M. ............................................. 13

Francke, L. ............................................. 42

Galazka, Z. ............................................. 32

Ganschow, S. ............................................. 32

Geer, R. ............................................. 81

Gupta, V. ............................................. 68

He, X. ............................................. 81

Henry, M. ............................................. 1

Huang, M. ............................................. 61, 101

Huso, J. ........................... 21, 48, 114

Huso, M. ............................................. 48

Iqbal, M. ............................................. 21

Ishizaki, H. ............................................. 120

Ito, S. ............................................. 120

Jagadish, C. ............................................. 75

Johnston, K. ............................................. 1

Klimm, D. ............................................. 32

Kobayashi, M. ............................................. 93

Kuznetsov, A. ............................................. 75

Kyndiah, A. ............................................. 13

Lacoe, R. ............................................. 53

Lambrecht, W. ............................................. 27

Limpijumnong, S. ............................................. 27

Matsui, H. ............................................. 87

McCluskey, M. ........................... 21, 48

McGlynn, E. ............................................. 1

McMaster, M. ............................................. 127

Minko, A. ............................................. 108

Moore, J. ............................................. 7

Morrison, J. ........................... 48, 114

Muller, H. ............................................. 53

Oder, T. ............................................. 127

Ono, Y. ............................................. 87

Prod'Homme, P. ............................................. 42

Rao, K. ............................................. 13

Rhodes, S. ............................................. 48

Rudenja, S. ............................................. 108

Sanchez, L. ............................................. 114

Schulz, D. ............................................. 32

Schwartz, R. ............................................. 53

Smith, A. ............................................. 127

Stansell, R. ............................................. 7

Sternagle, D. ............................................. 127

Svensson, B. ............................................. 75

Tabata, H. ............................................. 87

Tarun, M. ............................................. 21

Thapa, D. ............................................. 48

Tokranova, N. ............................................. 81

Tomar, M. ............................................. 68

T-Thienprasert, J. ............................................. 27

Tyagi, M. ............................................. 68

Uecker, R. ............................................. 32

# AUTHOR INDEX

Velpukonda, N. .................................... 127

Veuvonen, P. .......................................... 75

Vines, L. ............................................... 75

Voit, W. ................................................ 13

Wang, W. .............................................. 81

Wong-Leung, J. ..................................... 75

Wu, Y. .................................................. 13

Yaqoob, F. ...................................... 61, 101

Yeh, W. ................................................. 48

Young, H. ............................................ 114

Cambridge University Press
32 Avenue of the Americas, New York, NY 10013-2473, USA

Materials Research Society
506 Keystone Drive, Warrendale, PA 15086

ISBN 978-1-62748-214-1